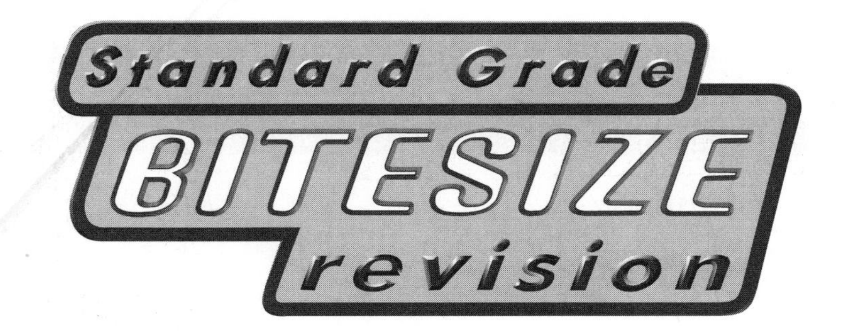

Susan Fraser

Published by BBC Educational Publishing,
BBC White City, 201 Wood Lane, London W12 7TS
First published 1999, reprinted 2000, 2001, 2002, 2003
© Susan Fraser/BBC Education 1999
BBC Educational Publishing would like to thank Graham Lawlor, author of GCSE Bitesize Revision: Maths, first published in 1998, for his contribution to the concept of this book.

ISBN: 0 563 46493 3
All rights reserved. No part of this publication may be reproduced, stored in any form or by any means mechanical, electronic, photocopying, recording or otherwise without prior permission of the publisher.

Designed by Steve Hollingshead
Printed in Great Britain by Bell & Bain Ltd., Glasgow

Contents

Introduction

About BITESIZE maths

BITESIZE maths is a revision guide which will help you with your **General Level Standard Grade** exams. You can record and then watch the TV programmes, work through this book and even get help on the Internet on-line service (see address at the bottom of this page). It's called BITESIZE maths because the best way to revise is in bite-sized chunks, not all in one go.

About this book

It's not possible for us to cover the whole of the **Standard Grade** maths course in this book and the TV programmes. Instead, we have decided to concentrate on topics at **General Level** which many students find difficult. The book is divided into six sections:

- Section 1: Statistics
- Section 2: Using number
- Section 3: Shape and space
- Section 4: Right angled triangles
- Section 5: Algebra
- Section 6: Using graphs

KEY TO SYMBOLS

A question to think about

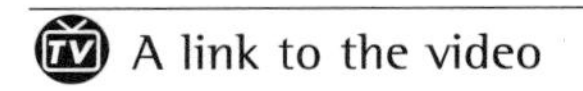

Each section has examples to show you how to answer questions, with some practice questions for you. There are also exam-style questions at the end of the book. Make sure that you understand the examples before you try these. You will also find FactZONES. These remind you of maths that you need to know well for the exam. If you don't understand something in a FactZONE, ask your teacher. Don't be embarrassed to ask for help – teachers like students to take responsibility for their own learning and their own future.

The book has lots of helpful hints and tips (REMEMBER! paragraphs in the margin), but space has also been left for you to make your own notes. If you have recorded the programme, note the time code from the video at the relevant place in the book, so you can quickly find that section again.

THE ON-LINE SERVICE

You can find extra support, tips and answers to your exam queries on the BITESIZE internet site. The address is http://www.bbc.co.uk/education/revision

About Standard Grade

Your school or college will enter you for your **Standard Grade** exam. There will be two written papers at each level, the first is non-calculator. Your marks come from Knowledge and Understanding (KU) and Reasoning and Enquiry (RE). With KU questions you can usually see which maths 'tool' you should use. RE questions make you think more about which type of maths to use. Your final exam result is decided equally from KU and RE.

Standard Grade exams are at three levels and everyone sits **General Level** and either **Foundation or Credit Level.** Your teacher will tell you which levels you are sitting, although you may know already.

Planning your revision

To develop any skill, you usually need to practise over and over until you get it right. When you were learning to walk, you didn't just stand up and start strolling along! You had to keep trying, practising taking steps again and again. Learning at school is much the same, especially in maths.

- The key to success is planning. Well before the exam you need to have a revision plan. Make a timetable. Put in the dates and what you will study. Most importantly, stick to your timetable and regularly review your progress. Ask for advice from your teacher.
- A very important part of maths is being able to use the basic maths 'tools'. For example, the tool might be how to work out a percentage or solve an equation. Don't just read the book, this doesn't work in maths. Try some practice questions for yourself. Take notes or draw diagrams and sketches – ask your teacher if you are unsure of which maths technique to use.
- Record the programmes from the TV and watch them as often as you need to. Using the video as well as your notes and drawings will help you to remember more easily.
- When you are going over a topic in class, make notes in your exercise book. After you have worked through some questions during the lesson, make some notes to remind you what to do.
- Probably the best preparation of all is to work through past exam questions. Ask your teacher or buy them from your local bookshop. Group the questions so that you do all of one type of question first, for example all of the Pythagoras questions. You will see that questions fall into a pattern – although the questions are not exactly the same, they do tend to be similar. Nearer the exam, do some complete papers and time yourself. You have 35 minutes for General Paper 1. Paper 2 takes 55 minutes. Use any time left at the end to check over what you have written.

Getting ready for the exam

- Make sure you know the date and time of your exam and where it is going to be held.
- You need a scientific calculator for Paper 2 of **General** and **Credit Levels**. Make sure you know how to use it. Try not to change calculators just before the exam, but if you have to use a different one, be sure you know how it works, especially for squares, square roots and trigonometry.
- Check your calculator and, if you think it needs new batteries, fit them. Make sure that your calculator is in degrees (shown as D or DEG on the display screen). If you see R, G, RAD or GRA, you will get some answers wrong. Find out how to put it back to degrees if it is in the wrong mode.
- When you use the calculator, try to have a rough idea of the size of answer you would expect. If you were doing 5.7 x 12, you would expect an answer of about 72 (6 x 12). If you got a very different answer, you would know you had done something wrong. Use your common sense.

On the day

- Arrive in good time, but not too early. Make sure you have a couple of pens and pencils, a rubber, ruler, and your calculator for Paper 2. When you go into the exam room, stay calm. Being slightly nervous is OK as the adrenaline will make you more alert. If you start to feel panicky, take some slow, deep breaths.
- Write in your name, school or college and candidate number on the front of your exam paper. The person supervising the exam, called the 'invigilator', knows this number if you have forgotten it. Then read the instructions carefully. The invigilator can answer general queries, but cannot help with the exam questions.
- At **General** and **Foundation Levels** you write your answers on the exam paper, at **Credit Level** you write in answer booklets of blank paper. You have to answer all the questions in the exam paper.
- When you are told you can start, take a couple of minutes to read through the questions. Answer the easiest ones first, then the more difficult ones. Remember to check that you haven't missed any pages or questions out.
- Look at the number of marks for each question, it will give you an idea of how much work you need to do. At **General** and **Credit Levels** you **MUST SHOW YOUR WORKING OUT** or you will not get all the marks for the question even if you are right. You don't need to show working out if the question says 'Write down', or if you have coordinates to plot, a drawing to do or information to read from a graph. At **Foundation Level** working out will get you more marks. Keep checking the time during the exam and if you are taking too long on a question, leave it and come back to it later if you have time. You will get marks for any correct working you do, even if you have not got to a final answer.
- Remember that you are like an athlete preparing for a big event – the more you prepare, the better you will perform on the day. Excellent preparation equals excellent results.

Good luck!

Acknowledgements

The author would like to thank her husband, Ian, and her parents, Sheila and Arthur Varley, for all their support and encouragement. Thanks also to Ruth Clark, Principal Teacher of Mathematics at Ullapool High School, and to her editor, Sarah Jenkin.

Statistics

This section is about:

- calculating mode, median and range
- constructing frequency tables
- constructing scattergraphs and drawing a best fitting straight line
- constructing and reading information from

  stem-and-leaf diagrams
- comparing two sets of data
- calculating probability

You will also need to study:

- calculating the average, or mean, on page 19
- the 'Using graphs' section on page 68

Statistics are used to predict what might happen. For example, keeping weather records helps weather forecasting, and knowing how many babies are born each year in a town or city helps to work out how many nursery school places will be needed. People sometimes keep statistical records for their own reasons.

Example 1

Neil likes to fish for brown trout. He keeps a record of the weight and length of each fish he catches.

Here is part of his record book.

Day	Weight	Length
Monday	123 g	49 cm
	22 g	25 cm
Tuesday	189 g	57 cm
	232 g	61 cm
Wednesday	206 g	59 cm
	22 g	26 cm
	148 g	52 cm
Thursday	–	–
Friday	244 g	61 cm
Saturday	273 g	62 cm
	65 g	39 cm

He uses his record book to work out several different statistics about the weight of the fish.

REMEMBER Writing the numbers in order, smallest first, is often the most useful first step in an answer.

First, he works out the **median** (middle) weight. He writes out the weights in order, from the lightest to the heaviest.

Weight in grams: 22, 22, 65, 123, 148, 189, 206, 232, 244, 273

When a list is in order, the median is the middle number. If there is an even number of entries in the list, as here, you add the two middle numbers together and divide by two.

The median weight = (148 + 189) ÷ 2 = 337 ÷ 2 = 168.5 grams

REMEMBER 'Modal' is just another way of asking you to 'find the mode'.

Next, Neil finds the **modal** weight. The mode is the number that occurs most often. 22 grams is listed twice, that is more often than any other number in the list, so the mode is 22 grams.

Then, Neil finds the **range** of the fish weights. The range is the highest number take away the lowest number. So in this list, the range is 273 - 22 = 251 grams

Finally, Neil works out the **mean** weight. The mean is the average, so he has to add up all the weights and divide the answer by how many weights there are.

REMEMBER To find out more about averages, look at page 19.

Mean = (22 + 22 + 65 + 123 + 148 + 189 + 206 + 232 + 244 + 273) ÷ 10
= 1524 ÷ 10
= 152.4 grams

Plot the two sets of data from Neil's record book – length and weight – onto a scattergraph. How closely related are weight and length?

Step 1: Draw the axes, one for length and the other for weight. Plot all of the lengths and weights just like a pair of coordinates (weight, length).

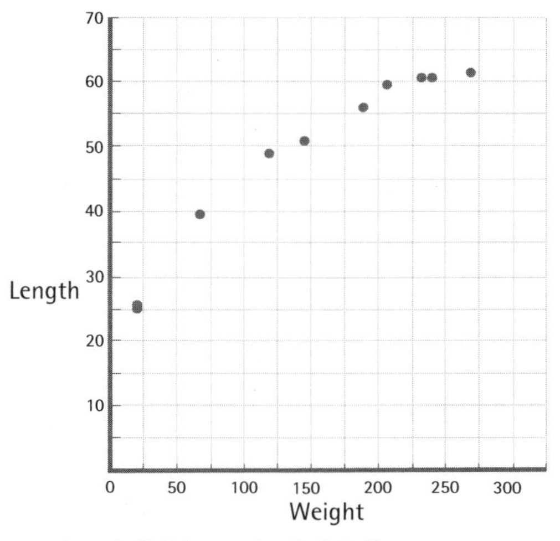

REMEMBER Plot the coordinates using dots not crosses.

Step 2: Draw a best fitting straight line.

Use a ruler to draw a line through at least two of the dots, so that about as many of the dots are on one side of the line as on the other.

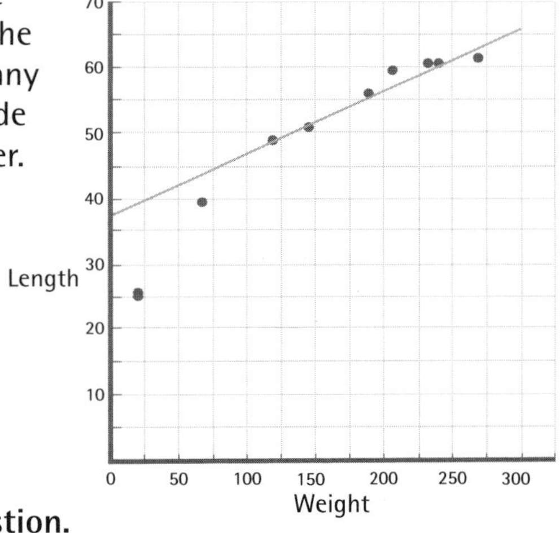

REMEMBER Your best fitting straight line can be different from someone else's but still right – there is more than one correct answer.

Step 3: Answer the question.

The graph shows that there is a strong relationship between the weight of the fish and their length because all of the dots closely follow a positive gradient line. This is called a positive correlation.

Example 2

Neil made this frequency table to show how many fish he caught each week.

Day	Tally	Frequency
Week 1	III	3
Week 2	~~IIII~~ I	6
Week 3	II	2
Week 4	~~IIII~~	5
Week 5	~~IIII~~ ~~IIII~~	10
	Total	26

REMEMBER You can also use the information on a frequency table to draw a bar graph (sometimes called a frequency diagram). To find out how to draw a bar graph, see page 73.

Another way to show this information is to use a stem-and-leaf diagram. Example 3 shows you how to draw a stem-and-leaf diagram.

Example 3

Here are the Blairbeg Youth Club attendance figures for one month already sorted into order.

No. of people attending: 6, 7, 8, 12, 14, 15, 15, 18, 20, 21, 22, 22, 23, 27, 29

To draw a stem-and-leaf diagram, start with the stem. In the stem, you write all of the number except the units figure. This is easy for most of these numbers, you put the tens digit into the stem.

But what happens to the numbers that don't have a tens digit, for example 6, 7 and 8?

You just write a zero to stand for the ten like this:

	0
Stem	1
	2

REMEMBER Write the unit digits next to the tens digits **in order,** for example 12, 14 and 15 would go next to the 1 in the tens line as 2 4 5.

Then write all the units alongside the correct number of tens in the stem.

0	6 7 8	
1	2 4 5 5 8	*Leaf*
2	0 1 2 2 3 7 9	

Write a key to explain the numbers: 0 | 6 stands for 6 people attending the Youth Club

So if the number was 154, 15 would be in the stem and 4 would be the leaf.

REMEMBER A stem-and-leaf diagram DOES NOT show when in the month people came to the Youth Club, because the numbers have been sorted into order!

Stem-and-leaf diagrams can be used to compare two sets of data. Here is the stem-and-leaf diagram showing the attendance figures at the Youth Club for two different months.

January		February
9 9 7 6 5 5	0	6 7 8
8 5 2 0	1	2 4 5 5 8
9 3	2	0 1 2 2 3 7 9

Key: 0 | 6 stands for 6 people attending the Youth Club

What does this diagram show you?

It shows that the attendance figures were lower in January than February. Most of the numbers in January belong to the stem that stands for units, February attendance was much better because most of the numbers are on the twenties stem.

Practice question

1) From Neil's record book, calculate the mean, median, mode and range using the **lengths** of the fish he caught.

Probability

Probability is a way of calculating how likely something is to happen.

If something is absolutely impossible, for example throwing a 7 on a normal six-sided dice, then its probability is 0 (zero).

If something is absolutely certain, such as there being a moon in the sky this month, then it has a probability of 1.

To calculate probability, you first need to work out how many possible outcomes or results there could be. For example, throwing a normal six-sided dice would give you 6 possible outcomes, picking a card from a pack would give you 52 possible outcomes.

Example 1

What is the probability of throwing a 4 with a normal six-sided dice?

To find the probability, you have to work out how often the winning outcome, called the favourable outcome, could occur.

There are 6 possible outcomes: you could throw a 1, 2, 3, 4, 5 or 6.

How many of these are a 4?

Only one is. So the probability of throwing a 4, written as P(4) is one chance in six, $\frac{1}{6}$ written as a fraction, or 0.17 as a decimal.

REMEMBER To change a fraction into a decimal, divide the top number in the fraction by the bottom number.

Example 2

In a first-year class there are 30 pupils. 12 of them are girls. What is the probability of picking a boy at random from this class?

Number of boys = 30 - 12 = 18

P(boy) = $\frac{18}{30} = \frac{3}{5}$ = 0.6

REMEMBER Probability is worked out on normal dice, playing cards, events, etc. Don't think 'But what if...' – probability doesn't work like that!

Practice questions

1) Calculate the probability of picking a King from a normal pack of playing cards.

2) What is the probability of picking a heart from a normal pack of playing cards?

Using number

This section is about:

- using the four basic rules of number
- changing fractions to decimals
- using percentages
- calculating money
- using ratio and proportion
- calculating time, distance and speed

The four basic rules of number are addition, subtraction, multiplication and division. When you are working on problems, you probably use words such as 'add', 'take away', 'times' and 'divide'.

In the exam you will need to work out which rule to use. You will also need to remember that addition and subtraction are opposites – they 'undo' each other. Multiplication and division are opposites, too.

When you are working with number it is important to have some idea of the size of answer that you expect, especially if you are using your calculator. Estimate your answer by rounding off the numbers and working out the sum. Look at your answer and ask yourself, 'Is this sensible?'

Many mistakes are made in working with time. You need to know facts about the number of days in the different months as well as days and weeks in a year.

Diagrams can help a lot in calculations.

For example, if you have a question about negative numbers, then drawing a number line, like this one, will help you to work out the answer.

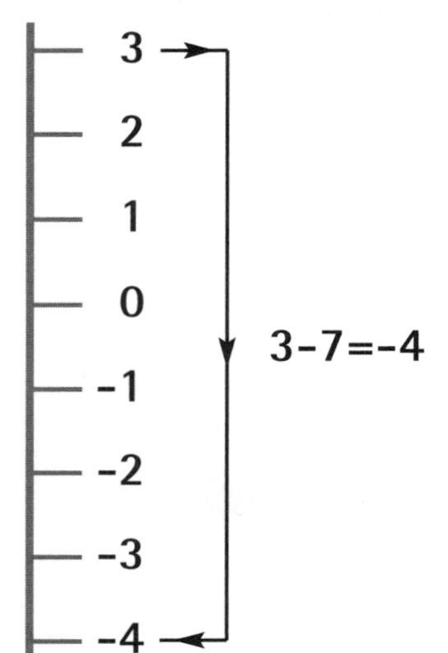

Use your common sense as well as your maths knowledge and you will get a better result.

FactZONE

Money

In most currencies the main unit is divided into 100 smaller units, e.g.

UK	£1	= 100 pence
Germany	1 mark	= 100 pfennig
USA	$1	= 100 cents

Interest

This is money paid on savings or loans. If you save money, you will be paid interest by the bank; if you borrow money, you pay interest to the bank.

Per annum means 'each year'.

Measurements

10 millimetres (mm) = 1 centimetre (cm)

1000 millimetres = 1 metre (m)

100 centimetres = 1 metre

1000 metres = 1 kilometre (km)

1000 grams (g) = 1 kilogram (kg)

1000 kilograms = 1 tonne

1 cm^3 = 1 millilitre (ml)

1000 cm^3 = 1 litre (l)

1000 millilitres (ml) = 1 litre

Time

60 seconds = 1 minute

60 minutes = 1 hour

24 hours = 1 day

7 days = 1 week

52 weeks = 1 year

365 days = 1 year

a leap year = 366 days (1 day more – this happens every fourth year)

(There is an easy way to remember the days in each month shown on page 19.)

The 24-hour clock:

It starts with midnight as 00.00

1 am is 01.00 and 1 pm is 13.00

So 20.15 is 8.15 pm and 07.55 is 7.55 am

Percentages

Per cent means out of 100

Common percentages as fractions:

$1\% = \frac{1}{100}$ $10\% = \frac{1}{10}$ $20\% = \frac{2}{10}$

$25\% = \frac{1}{4}$ $50\% = \frac{1}{2}$ $75\% = \frac{3}{4}$

$33\frac{1}{3}\% = \frac{1}{3}$ $66\frac{2}{3}\% = \frac{2}{3}$

100% is one whole

Using number to solve problems

Making sense of number

REMEMBER You should try answering the questions before looking at the method given.

Ann sells electrical equipment packed in boxes.

Each box is 55 cm high.

Boxes must not be stacked more than 2.75 metres high.

How many boxes can she safely stack?

Look carefully at the question. What are you being asked to do?

Step 1: Write down what you know.

Total height = 2.75 metres

1 box = 55 centimetres

Step 2: Decide on the maths you need to use.

You can't do anything with different units of measurement, so make them both the same:

Total height = 275 centimetres

1 box = 55 centimetres

Now decide what to do with the numbers.

Here you need to divide:

$275 \div 55 = 5$

Step 3: Make sense of the answer.

5 boxes can be stacked on top of each other.

How can you check if this is right?

You can work backwards:

5 x 55 = 275 centimetres = 2.75 metres

REMEMBER In an exam you often have to take the question apart and decide what kind of maths you need to do.

Make things easier

If you find that the numbers confuse you, give yourself easier ones. For example, make the total height 6 metres and a box 2 metres high.

Work it out so that the answer makes sense: $6 \div 2 = 3$ boxes high.

Then do the exactly the same with the numbers in the question.

Insurance and hire purchase

The cost of insurance depends on what you insure and where you live. It costs more in a city such as Aberdeen than in more remote areas such as Skye or Wick.

	Contents cost per £100	Building cost per £1000
Area 1	£1.65	£1.82
Area 2	£1.42	£1.55
Area 3	£1.18	£1.30

How much would it cost altogether to insure a house (building) for £80 000 and its contents for £1800 in Area 2? (See table on the right.)

Step 1: Work out the building insurance.

The building insurance is per £1000. You pay £1.55 in Area 2 for each £1000 of insurance cover.

First work out how many £1000 there are in £80 000: 80 000 ÷ 1000 = 80

Now multiply the cost by 80: £1.55 x 80 = £124

Step 2: Work out the contents insurance.

The building insurance is per £100. You pay £1.42 in Area 2 for each £100 of insurance cover.

First work out how many £100 there are in £1800: £1800 ÷ 100 = 18

Now multiply the cost by 18: £1.42 x 18 = £25.56

Step 3: Add the two amounts together to get the total cost for insurance.

£124 + £25.56 = £149.56

REMEMBER Any information you need in the exam question is given in a table like the one above.

REMEMBER Look at your answer and check that it makes sense!

Hire purchase

Hire purchase lets you buy things, such as a car, by borrowing money and paying it back over a period of time. You usually have to pay some money when you buy the item. This is called the deposit and is often worked out as a percentage of the total price.

Margaret bought a car. It cost £5995.00. She had to pay a 15% deposit and then pay 36 monthly payments of £182.34.

How much extra did she pay?

Step 1: Work out the deposit.

15% of £5995: 5995 x 15 ÷ 100 = £899.25

Step 2: Work out the monthly payment total.

36 x £182.34 = £6564.24

Step 3: Add the two amounts together to find the total amount she paid.

£899.25 + £6564.24 = £7463.49

Step 4: Subtract the cost price of the car to work out the difference.

£7463.49 - £5995 = £1468.49 extra

Fractions, decimals and percentages

Fractions, decimals and percentages appear every day in our lives, not just in your exam! You need to know how to handle them successfully.

Fractions

- The number on the top is called the **numerator**.
- The number on the bottom is called the **denominator**.

To work out the fraction of an amount:

Divide the amount by the denominator (bottom number) and multiply by the numerator (top number).

For example, to find $\frac{4}{5}$ of £90 work out:

90 ÷ 5 = 18

then 18 x 4 = £72

(or on your calculator 90 ÷ 5 x 4 = 72 all in one step).

Scientific calculators can work out fractions. Look for the fraction key, which looks like this $\boxed{a\frac{b}{c}}$

If $\boxed{a\frac{b}{c}}$ is written on the key, you just press that key to use it. If it is written above the key, you will need to press the '2nd F', 'SHIFT' or 'INV' key first, then the fraction key.

The fraction key is used to separate the 'parts' of the fraction from each other. For example, to work out $1\frac{1}{2} + 4\frac{1}{4}$ on the calculator, enter:

1 ❑ 1 ❑ 2 + 4 ❑ 1 ❑ 4 = (❑ means press the fraction key here)

On the display screen you will get something like:

5 ┌ 3 ┌ 4

This is read as $5\frac{3}{4}$

You can use fractions in calculations in exactly the same way as you would use whole numbers.

To change a fraction to a decimal:

Divide the numerator by the denominator.

For example, $\frac{3}{4}$ = 3 ÷ 4 = 0.75

REMEMBER You need to estimate mentally the answer, so that you know if you have used the calculator properly.

Percentages

To change from a percentage to a decimal:

Divide the percentage by 100.

For example, to show 30% as a decimal:

$\frac{30}{100} = 0.3$

To change from a fraction to a percentage:

Multiply the fraction by 100.

For example, to show $\frac{3}{4}$ as a percentage:

$\frac{3}{4} \times 100 = 75\%$

To find a percentage of a quantity:

Multiply the amount by the percentage and then divide by 100.

For example, to find 45% of 1080:

$1080 \times 45 \div 100 = 486$

The answer is given in the same type of quantity. So if the amount was £1080, the answer would be £486, if it was 1080 sheep, the answer would be 486 sheep, and so on.

To find the percentage increase (or surcharge):

Work out the percentage and add the amount on.

For example, if it is an 8% surcharge on a holiday costing £164.00:

8% of 164 = $164 \times 8 \div 100$ = £13.12

The holiday now costs 164 + 13.12 = £177.12

To find out the percentage decrease (or discount):

Work out the percentage and subtract it from the original amount.

Example 1

Calculate a 15% discount on a television costing £299.00:

15% of 299 = $299 \times 15 \div 100 = 44.85$

The television now costs 299 - 44.85 = £254.15

Example 2

A typical Knowledge and Understanding question would be as follows:

Simon invests £540 in an account with the Inverness Building Society at 3.8% per annum.

Calculate the interest he should receive after 4 months.

REMEMBER
For pounds and pence you must use a £ sign and two numbers after the decimal point.

REMEMBER
It is a good idea to give your answer to at least one decimal point, unless the question asks you otherwise.

REMEMBER

On a calculator:

4.3	means £4.30p
4.03	means £4.03p

Step 1: Using the percentage given, calculate the interest for a year.

540 x 3.8 ÷ 100 = £20.52

Step 2: Use your knowledge about time, i.e. the number of months in a year.

The money is only in the Building Society for 4 months. To work out 1 month's interest, divide the amount of interest by 12. Then multiply by 4 to work out the interest for 4 months:

20.52 ÷ 12 x 4 = £6.84

Example 3

A Reasoning and Applications question is different to a Knowledge and Understanding question, as it makes you think more about which type of maths to use. A typical Reasoning and Applications question would be as follows.

A fruit pie has to contain between 75% and 80% of fruit.
A 'Tasty' fruit pie weighing 950 grams is found to have 693 grams of fruit.

◎ *Does the 'Tasty' fruit pie meet the requirement? Give a reason for your answer.*

Step 1: Work out what percentage of fruit is in the pie.

To work out one amount as a percentage of another, divide the amount (here the weight of fruit) by the total amount (here the weight of the whole pie) and multiply it by 100.

REMEMBER

Give all the information needed – yes or no and a reason based on your working out and the original question.

pie = 950 grams

fruit in the pie = 693 grams

693 ÷ 950 x 100 = 72.94% = 72.9%

Step 2: Give your answer. Remember to give a reason, as the question asks.

No, it doesn't meet the requirement because it has 72.9% fruit, which is less than 75%.

Practice questions

1) Work out $\frac{5}{8}$ of £64.

2) Ross invests £19 000 in a bank account which pays 3.5% per annum.

   He invests the money for 8 months.

   Calculate the interest he earns on his money.

3) Mary gets a 3% pay rise.

   If she earned £115.80 before the pay rise, what will her new pay be?

Averages

Sometimes questions look like one thing and are really something else.

The following question looks like a percentage question, but isn't. It is actually testing your knowledge of averages.

A teacher tested her class and worked out the average mark for the 16 pupils. She found that the average was 54%.

She noticed that one pupil had a mark of 99% – much higher than anyone else. She decided to work out a new average, leaving the highest mark out.

◎ *What was the new average?*

? *How do you work out the average of a group of numbers?*

Add up all the numbers and divide the total by how many numbers there are. For example, you can work out the average of 6, 7, 3, 5, 8, 2, 8, 9 by adding all the numbers together to get 48, then dividing by 8 (how many numbers there are) to get the answer 6.

The question above is asking you to use the information in a different way.

You know the average: 54%.

You know how many numbers there were to begin with: 16 pupils.

Step 1: Work backwards to find out the total mark for everyone.

54 x 16 = 864

Step 2: Take away the highest mark.

864 - 99 = 765

Step 3: As this is now the total for 15 pupils, divide the new total by 15.

765 ÷ 15 = 51%

So 51% is the new average mark.

REMEMBER Even though you are only adding, subtracting and dividing numbers, you must show this in your working out. **Remember no working = no marks!**

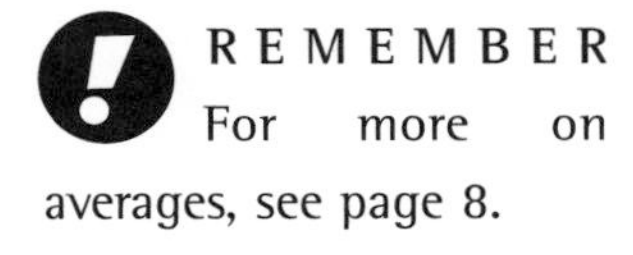

REMEMBER For more on averages, see page 8.

Practice questions

1) Work out the average of the test marks for this class:

   63, 42, 97, 56, 51, 43, 81, 65, 69, 77, 61, 89

   Give your answer to the nearest whole number

2) The average weight of a squad of 14 football players was 63.4 kg.

   Another person joined the squad. The average weight was now 64.12 kg.

   What did the new team member weigh?

Foreign exchange

If you were going on holiday, you would need to change your British money into the currency (money) of the country you were going to visit. This is called foreign exchange.

In Britain the exchange rate is always given as the amount of foreign currency you can get for £1.

This is a typical foreign exchange rate table:

German mark	2.74	Spanish pesetas	232.40
Italian lire	2731	US dollars	1.61
Portuguese escudos	280.14	Euros	0.7

From this list you can see that you can get 2731 lire or $1.61 for £1.

To change British pounds into a foreign currency, you multiply by the exchange rate.

For example, £90 in German marks:

90 x 2.74 = 246.6 marks

To change foreign currency back to pounds, you divide by the exchange rate.

For example, 15 400 lire in £:

15 400 ÷ 2731 = £5.639 = £5.64 – not a lot when you thought you had thousands!

How much money do you get each week? Look in the paper or on TV for the exchange rate and work out how much you would get in Euros.

Practice questions

1) Ruaraidh went on holiday to the USA.

   Before going on holiday, he changed £450 into US dollars.

   The rate of exchange was $1.61 to the £. How many dollars did he get?

2) Sara returns from holiday with 80 Dutch guilders.

   The exchange rate is 3.2 guilders to the £.

   The bank charges £2 for changing money.

   How much British money does she get back at the bank?

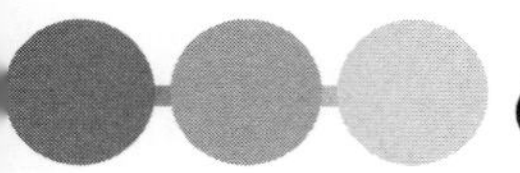

Ratio and proportion

Ratio is a way of describing the sharing out of something.

Example 1

You have £30 and you want to give 1 share to James and 2 shares to Barbara.

How many shares are you giving out?

You would be giving out 3 shares altogether.

So now how much money would James and Barbara each get?

Each share would be £30 ÷ 3 = £10.

So James (1 share) would get 1 x £10 = £10.

Barbara (2 shares) would get 2 x £10 = £20.

The ratio of James's share to Barbara's share is 1 share to 2 shares or 1:2.

REMEMBER You can check your answer. The total amount that James and Barbara get altogether is £10 + £20 = £30. That is the amount you started with.

Example 2

A wealthy farmer leaves £49 000 to her two children in the ratio 2:5.

How much does each child get?

How many shares are to be given out?

Step 1: Work out how much one share is.

The ratio is 2:5, so that is 2 shares to 5 shares, making 7 shares altogether.

One share is £49 000 ÷ 7 = £7000

Step 2: Work out how much money each child gets.

The first child gets 2 shares of £7000 = 2 x 7000 = £14 000

The second child gets 5 shares of £7000 = 5 x 7000 = £35 000

Step 3: Check your answer.

£14 000 + £35 000 = £49 000 (the total shared out)

REMEMBER To answer this sort of question you must use your common sense as well as your maths knowledge.

Example 3

If it takes 3 painters 6 days to completely paint a new house, how long will it take 4 painters?

You can work out how long it will take by setting out your working like this:

	3 painters	need	6 days
	÷ 3		**x 3**
	1 painter	needs	6 x 3 = 18 days
	x4		**÷ 4**
so	4 painters	need	$(18 \div 4) = 4\frac{1}{2}$ days

REMEMBER Use your common sense. Is it going to take more or less time if more painters work on the job? Of course, it will take less time!

Time

12-hour time is written using **am** to show that it is before midday and **pm** to show that it is after midday.

24-hour time starts counting the time from midnight, written as 00.00 and counts on through the day: 01.00 = 1 am, 07.45 = 7.45 am and so on. When the time has moved to 1 pm, this is shown as 13.00 (1 + 12 = 13). So 5.20 pm is 17.20 (5 + 12 = 17).

Changing back to 12-hour time is easy, 20.55 is 8.55 pm (20 - 12 = 8) and 23.05 is 11.05 pm (23 - 12 = 11). See the FactZONE on page 13.

To change from 12-hour to 24-hour time, add on 12 to the hours if the time is 1 pm or more. To change from 24-hour to 12-hour time, do the opposite, i.e. subtract 12 from the hours if the hours are 13 or more.

REMEMBER Write am after the time if it is before midday and pm if it is after midday, but *only* for 12-hour time.

Calculating time

The easiest way to calculate time is to use the counting-on method. Don't try to use your calculator – it doesn't work!

To work out the time from 10.50 am to 5.35 pm use a diagram.

Break the time down into minutes and hour-long jumps starting and ending at the times you need.

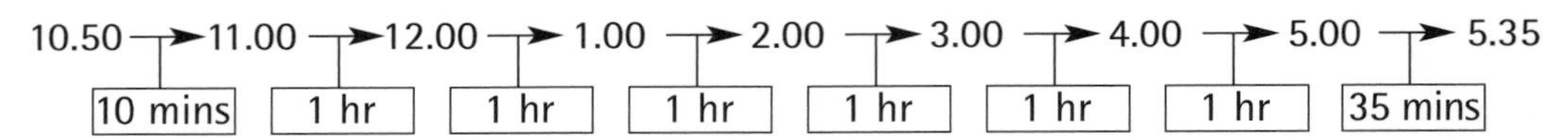

Then add up the number of minutes and the number of hours.

= 10 minutes + 6 hours + 35 minutes = 6 hours 45 minutes

You then have the right answer.

Remember that 60 minutes = 1 hour. If the number of minutes is more than 60, subtract 60 from your number of minutes and add on one more hour.

Example

Sarah started work at 7.05 am and finished at 11.45 am.

How long did she work?

Step 1: Break the time down into minutes and hour-long jumps.

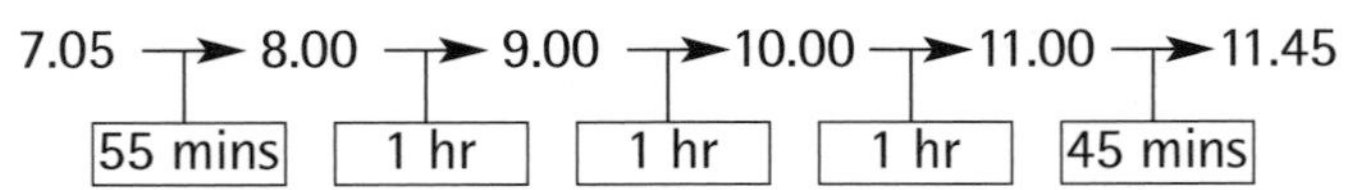

Step 2: Add up the minutes and hours.

55 minutes + 3 hrs + 45 minutes = 3 hours 100 minutes

Step 3: As the minutes add up to more than 60, work out how many extra hours there are.

100 - 60 = 40 minutes

3 + 1 = 4 hours

So the time is 4 hours 40 minutes.

To change hours in decimals to hours and minutes:

You might be given a time written as a decimal and be asked to turn that time into hours and minutes. To do this you write the whole number part as hours. Then multiply the decimal part by 60 and write that down as minutes.

For example, 7.2 hours is 7 hours and 0.2 x 60 = 12 minutes

so 7.2 hours = 7 hours and 12 minutes.

To change hours and minutes to hours in decimals:

To do this divide the minutes by 60 and add the hours.

For example, 10 hours 48 minutes:

minutes 48 ÷ 60 = 0.8

0.8 + 10 = 10.8 hours.

Days in the months of the year

Here is an easy way to remember which months have 31 days. Make fists and look at the backs of your hands. Count off the months like this:

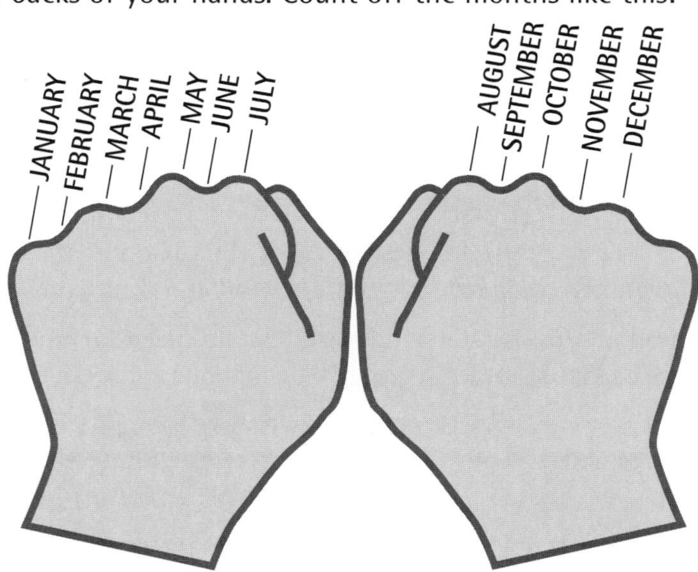

The 31-day months are always on a knuckle.

All the 30-day months are in a dip in between the knuckles.

You will have to remember that February is different – it has 28 days normally and 29 in a leap year.

REMEMBER

You can use this rhyme to remember the number of days in each of the months of the year: '30 days hath September, April, June and November. All the rest have 31, except February alone which has 28 days clear and 29 days each leap year.'

Time, distance and speed

REMEMBER Units of measurement match up. Miles and hours give you miles per hour. Metres and seconds give you metres per second.

The DST triangle will help you to remember what to do in problems using time, distance or speed.

Draw the DST triangle and write in the letters.

Circle the one you have to find and the triangle will tell you what to do with the other two.

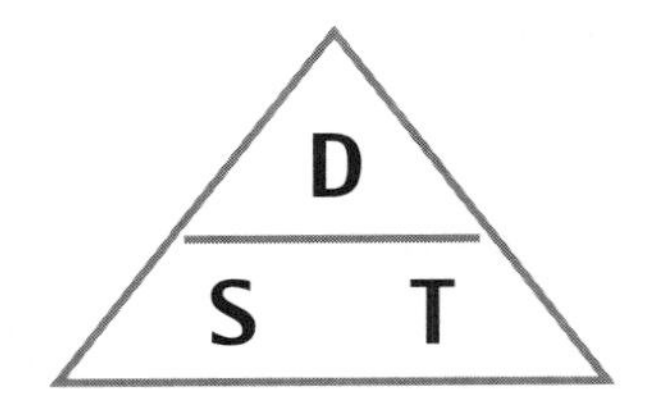

Example

A train was travelling at 85 miles per hour for 2.3 hours.

How far did it travel?

REMEMBER In algebra: ST means S x T and $\frac{D}{T}$ means D ÷ T

You need to find the **distance**.

Draw the triangle and circle the D.

You are left with S and T on the bottom line.

So you multiply speed and time:

85 x 2.3 = 195.5 miles

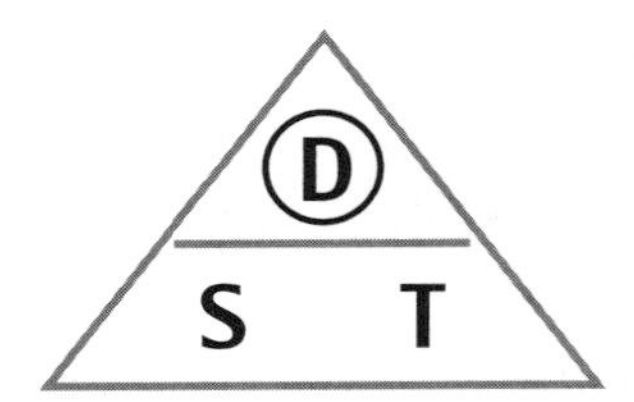

To work out the **speed** in a problem, circle the S.

You are left with D on the top line and T on the bottom line.

So you divide distance by time.

To work out the **time** in a problem, circle the T.

You are left with D on the top line and S on the bottom line.

So you divide distance by speed.

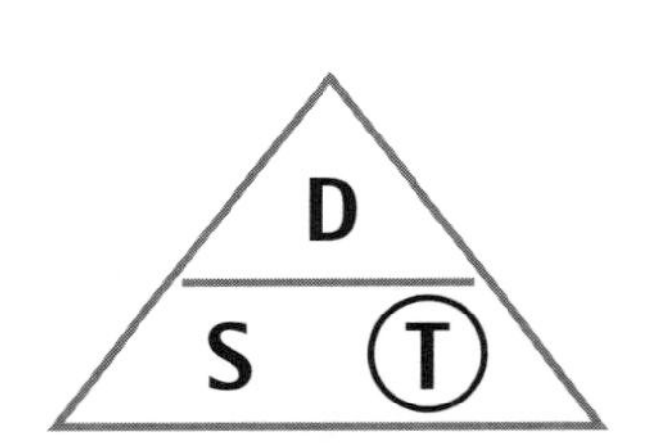

Remember, if you are given hours and minutes in a question, you will need to change the time to hours in the form of a decimal (see page 23).

Practice questions

1) A car travels at a steady speed of 65 kilometres per hour.

   How far did it travel in $4\frac{1}{2}$ hours?

2) How long does it take to fly from Inverness to Luton Airport, London, a distance of 560 miles at a speed of 452 miles per hour? Give your answer in hours and minutes.

Reading two-way tables

You have to be able to read two-way tables and use the information contained in the tables to solve problems.

Aberdeen					
125	Edinburgh				
165	146	Fort William			
145	44	104	Glasgow		
105	158	66	165	Inverness	
80	45	105	62	113	Perth

This table gives distances in miles between some of the towns and cities in Scotland. To find the distance between two places, find their names on the table and follow the lines across and down until they meet.

For example, the distance from Edinburgh to Inverness:

125	Edinburgh				
165	146				
145	44	104			
105	158	66	165	Inverness	
80	45	105	62	113	

= 158 miles

REMEMBER You can draw lines like these on your exam paper to help you find the answer.

You will often be asked to use this information in a time, distance and speed question.

Practice questions

1) A train leaves Aberdeen at 11.54 and arrives in Inverness at 14.18.

   How long did it take?

2) Look at the table above. How far is it from Aberdeen to Inverness?

   What was the speed of the train in question 1?

Shape and space

This section is about:

- using bearings
- using scale drawings
- working with angles
- finding the area and circumference of a circle
- calculating the area of a rectangle, triangle, kite, rhombus or a composite shape
- working with volumes of a cuboid or a prism
- working with reflection and rotation symmetry
- using coordinates

Before you begin this section, make sure that you understand the basic measurements given in the FactZONE on page 13.

At the beginning of the **General Level** exam paper, you are given a formulae list. The formulae that are covered in this unit are:

Circumference of a circle: $C = \pi d$

Area of a circle: $A = \pi r^2$

Curved surface area of a cylinder: $A = 2 \pi r h$

Volume of a cylinder: $V = \pi r^2 h$

Volume of a triangular prism: $V = A h$

This means that you do not need to remember these, you just need to know how to use them. They will be covered in detail as you work through this section.

However, there are a number of other facts that you will have to memorise, mostly to do with metric measurements (see the FactZONE on page 13) and angles (see the FactZONE opposite).

Make sure that you know your angle and circle facts well – you need to know the names of the parts of a circle.

You should understand the meaning of squaring:

d^2 means d x d and **not** 2 x d

$A = \pi r^2 = A = \pi \times r \times r$

You will need to know how to round numbers up and down, to a number of decimal places or to a significant figure or to the nearest unit of 10. You also need to practise rounding to a sensible degree of accuracy. Ask your teacher if you are unsure about rounding.

FactZONE

Basic angle facts

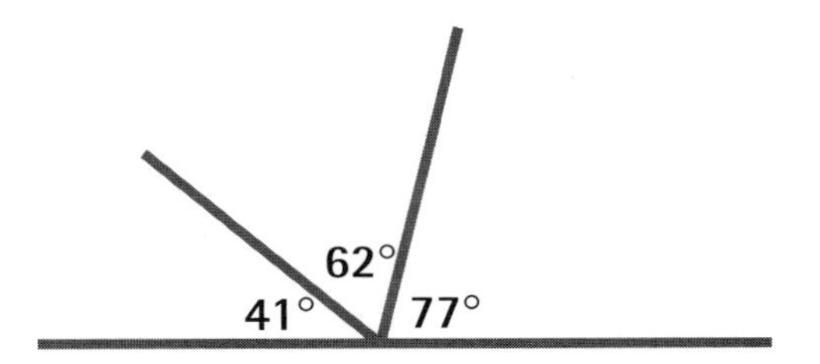

The angles on a straight line add up to 180°.

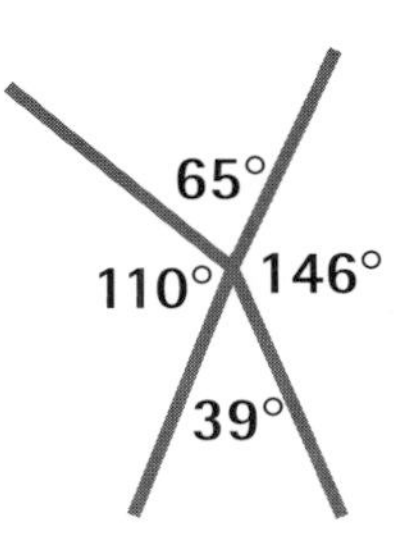

The angles around a point add up to 360°.

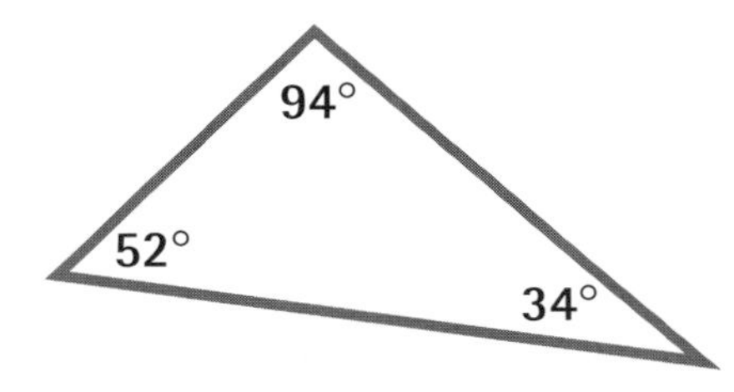

The angles in a triangle add up to 180°.

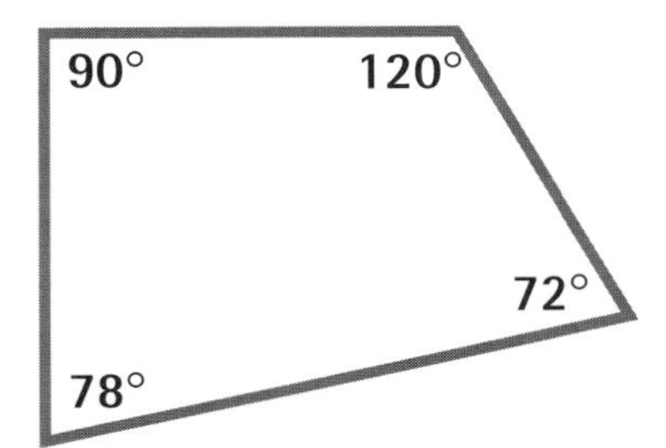

The angles in a quadrilateral add up to 360°.

Angles and lines

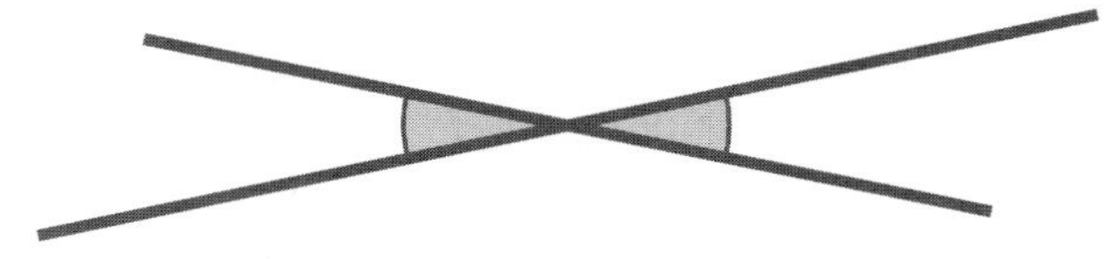

Opposite angles, sometimes called **X angles**, are equal.

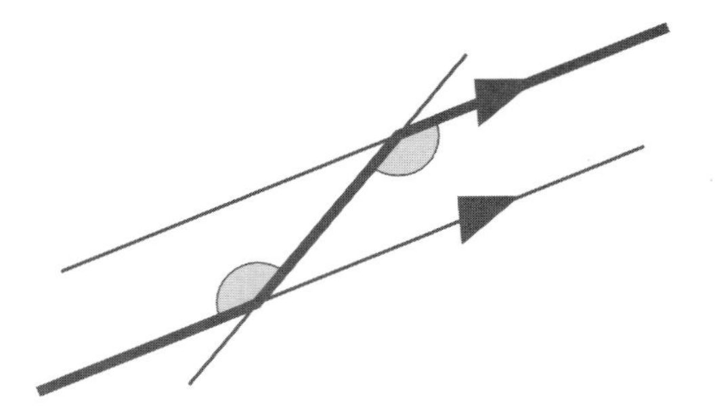

Alternate angles, often called **Z angles**, are equal.

The 'Z' shape can be back to front or stretched out (as it is here).

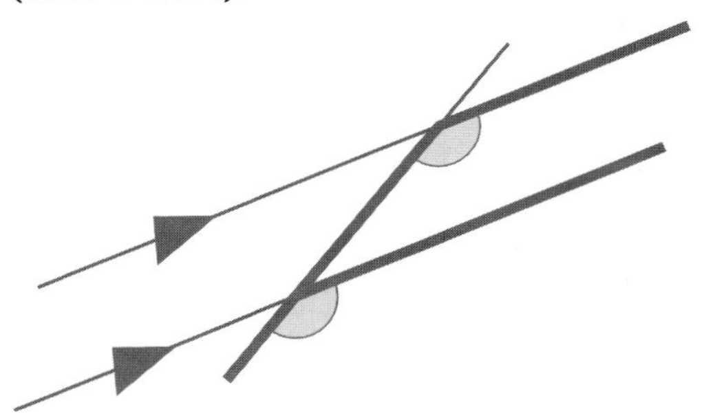

Corresponding angles, sometimes called **F angles**, are equal.

The 'F' shape can be back to front or upside down.

Shapes

A quadrilateral is a four-sided shape. A square, a rectangle, a rhombus and a kite are all quadrilaterals. A pentagon has five sides. A hexagon has six sides.

If a shape is called 'regular', for example a 'regular hexagon', then all of its sides and angles are the same size.

Bearings

Bearings are used to give a direction in which to find something, for example a plane flying from Glasgow to New York or a helicopter flying to an oil rig at sea.

There are two key points to remember:

- bearings are always measured in a clockwise direction from a north line
- bearings are always written with three digits.

REMEMBER Clockwise means turning the same way that the hands of a clock turn.

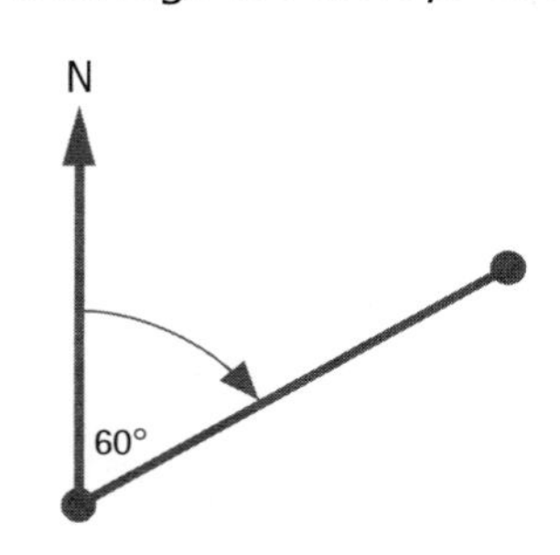

The angle between the two lines is 60°.

So the bearing is 060°.

Angles with three digits are still written in the same way, for example bearing 125° is still written as 125°.

Example

The bearing of the Bravo Oil Rig from Aberdeen Airport is 060°.

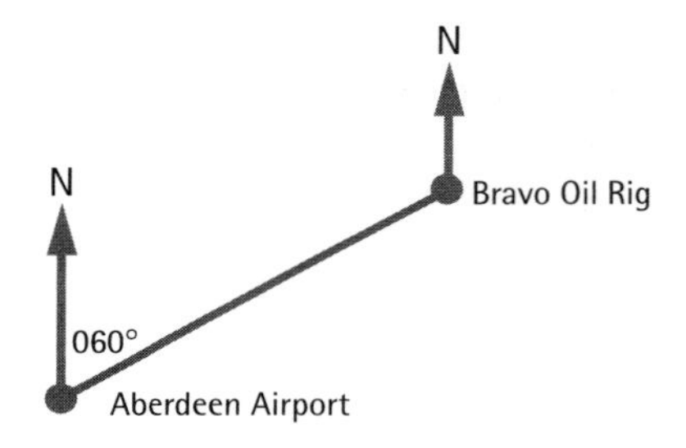

What is the bearing of Aberdeen Airport from the Bravo Oil Rig?

Step 1: Calculate the marked angle from the north line at the oil rig.

Step 2: Extend the north line down to give you a 180° angle and a smaller one.

The smaller angle is 60°, because the two north lines are parallel.

Step 3: Add the large and the small angles together.

So the bearing is 180° + 60°
= 240°

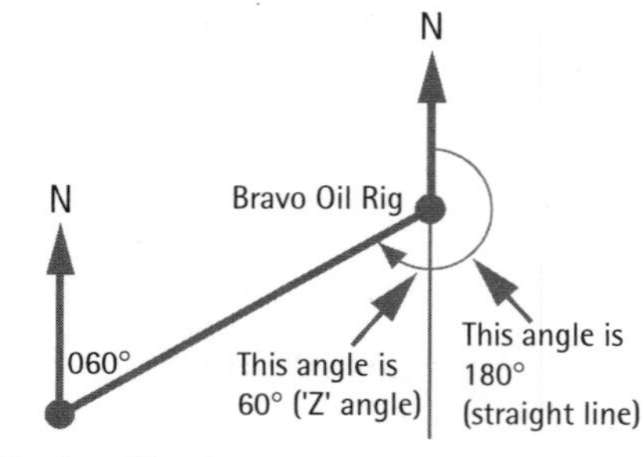

Compass points and angles

A question on bearings might include information about a point of the compass and ask you to work out a bearing.

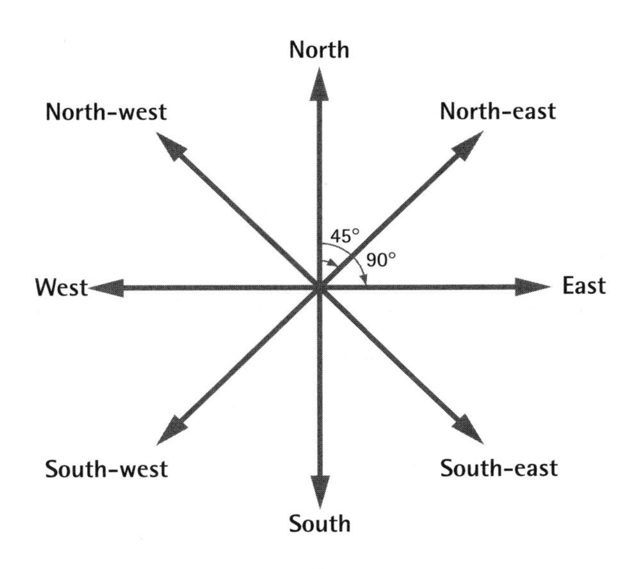

The main compass points are north, east, south and west.

Between each of the main compass points, there is a right angle (90°).

The compass points north-east, south-east, south-west and north-west are half way between the main compass points, so the size of the angle between these is 45°.

So if you are told that Q is due east of P, it tells you that the angle NPQ is 90°.

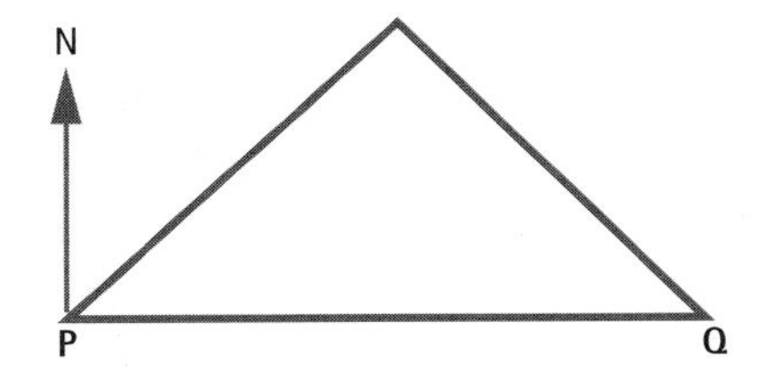

Practice question

1) The sketch (not drawn to scale) shows the position of two fishing boats, A and B, and a coast guard station, marked C.

Fishing boat B is 105 km due east of the lighthouse at C.

Fishing boat A is 65 km from C and on a bearing of 30°.

a) What is the size of angle ACB?

Now read pages 30 and 31 before you answer b) to d).

b) Make a scale drawing of the position of the lighthouse and fishing boats.

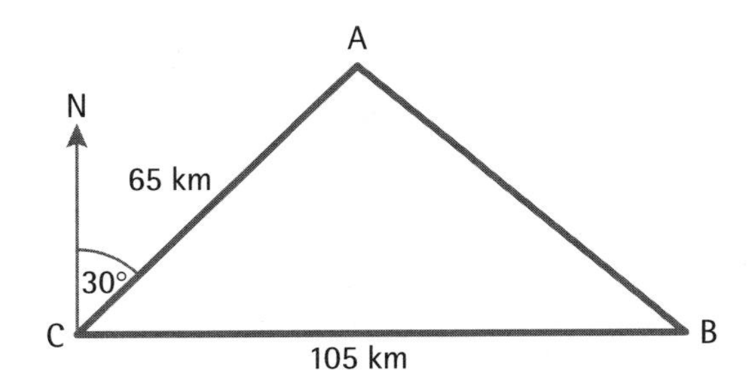

c) Use your scale drawing to find the size of angle ABC.

d) Calculate the bearing of A from B.

Scale drawing

> **REMEMBER** Before you begin to work through this section, make sure that you understand about bearings (page 28).

Scale

Sometimes in an exam question you will be given the scale you have to work to. Sometimes you have to decide your own.

A scale could be shown like this: Scale: 1 cm to 10 km

This means that every 1 centimetre you measure on the page is actually 10 kilometres in real life.

So if the real distance is 35 km, you must divide 35 by 10 to find out how many centimetres to draw: $35 \div 10$ gives 3.5 cm on your drawing.

If you measure 6.7 cm on your drawing, to find out the real-life distance you must multiply by 10: 6.7 cm x 10 = 67 km in real life.

If you have to decide the scale you will use, look at the real-life sizes you have to work with.

Make your scale easy by using round numbers, for example 1 cm to 2 km, 1 cm to 5 km or 1 cm to 20 km.

Don't choose a scale that will make your drawing too small, as that will make it difficult to measure accurately.

> **REMEMBER** When you have decided what scale to use, remember to write down that scale on your drawing.

Example

Some walkers start a trek from their overnight camp. They walk on a bearing of 104° for 5 km. They stop for a rest. They then walk north-east for 8 km.

- *Use a scale of 1 cm to 1 km to make a scale drawing of their walk. The beginning of the scale drawing has been started below.*
- *Use your scale drawing to find (i) the bearing back to their camp (ii) the distance back to their camp.*

Step 1: Change the distances given in the question to measurements for your page, using the scale in the question.

The first distance is 5 km. This is 5 cm on a page. The second distance is 8 km. This is 8 cm on your page.

Step 2: Using the north line, start your drawing with the first distance.

Start from the camp site.

Use a protractor to measure 104° from the north line. Draw your line 5 cm long.

Draw another north line where your pencil stops.

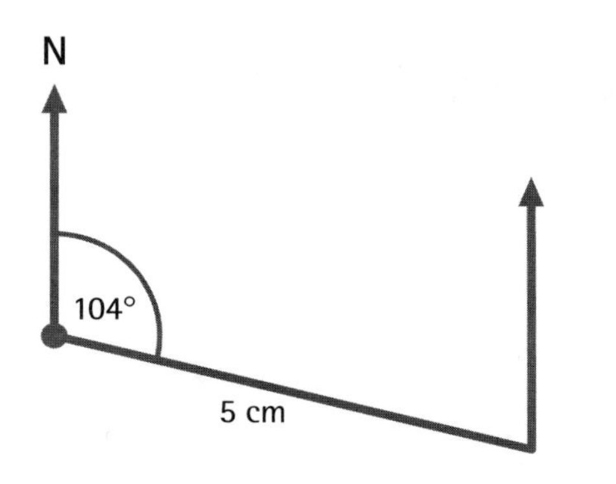

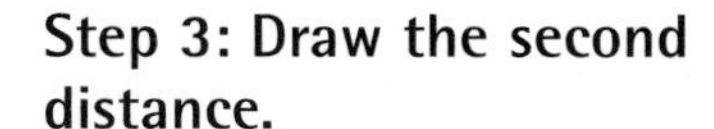

Step 3: Draw the second distance.

Draw in another line,
8 cm long going north-east.
North-east is halfway
between north and east,
a bearing of 045°.

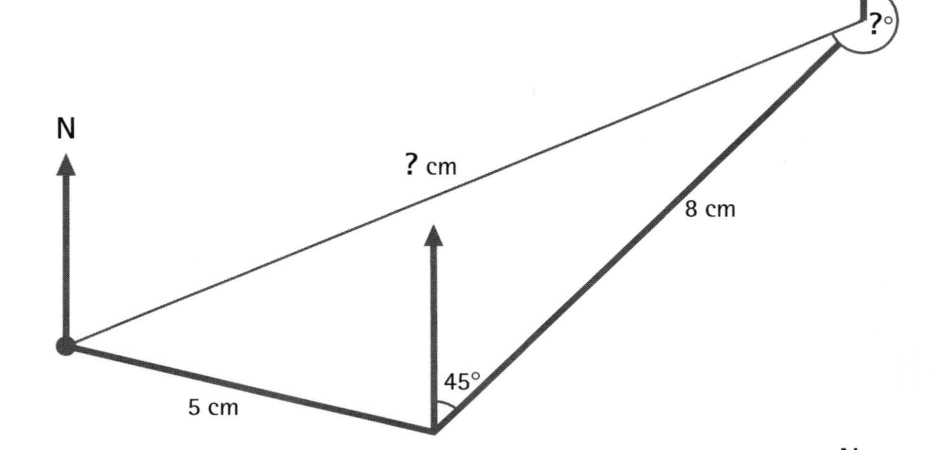

Step 4: Find the bearing back to their camp (the angle marked ?°).

If you have an angle measurer (360°), you will have no difficulty.
The bearing is 246°.

If you use a protractor (180°), you will need to extend the north line so that you can measure the smaller (shaded) angle. It is 66°.

The larger (striped) angle is 180°.
So the bearing is 66° + 180° = 246°.

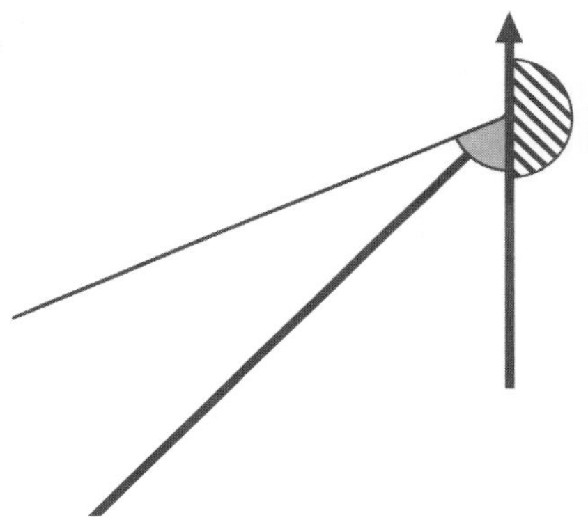

Step 5: Find the distance back to their camp (line marked ? cm).

Use a ruler to measure the distance marked ? cm on the drawing. It is 11.4 cm.

Using a scale of 1 cm to 1 km, this makes the distance back to their camp 11.4 km.

Shape facts

REMEMBER Don't measure an angle unless you are told to. Often the question will say 'calculate', which means 'work out the answer'. You should use the information given in the question and the angle facts from this section.

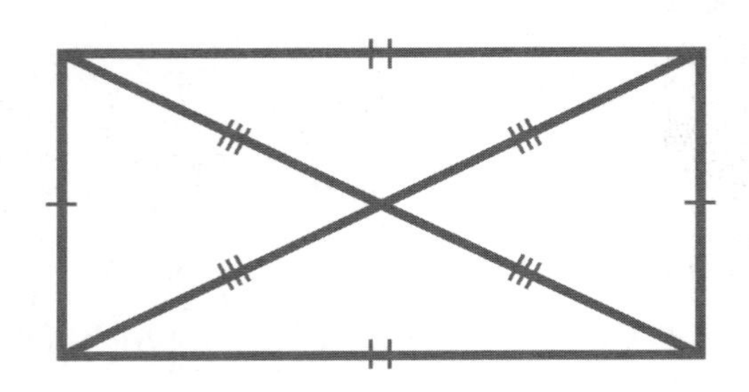

Rectangle

- opposite sides are equal and parallel
- sides meet at a 90° angle
- the diagonals are equal in length
- the diagonals bisect each other
- opposite angles at the centre are equal

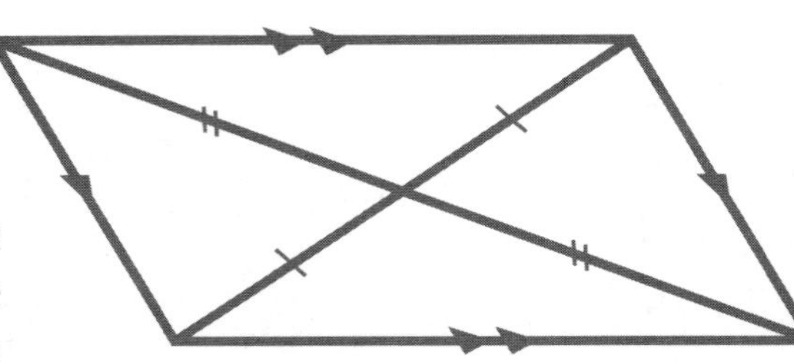

Parallelogram

- opposite sides are equal and parallel
- opposite angles are equal
- the diagonals bisect each other
- opposite angles at the centre are equal

REMEMBER The lines that are equal in length are shown here with an equal number of small dashes on them. The lines with the same number of arrows on them are parallel.

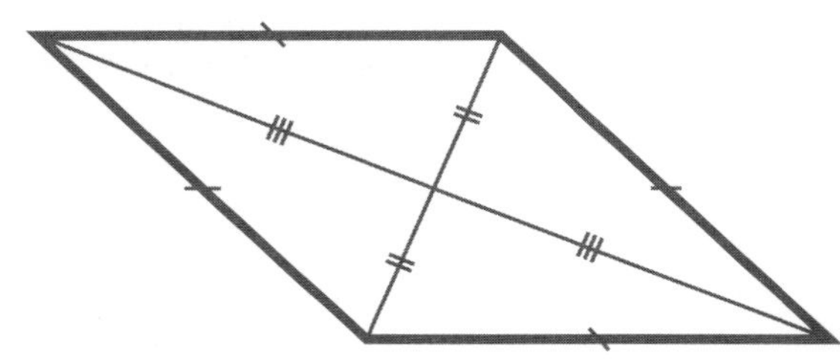

Rhombus

- all sides are equal
- the diagonals meet at a 90° angle
- opposite angles are equal
- the diagonals bisect each other
- opposite angles at the centre are equal

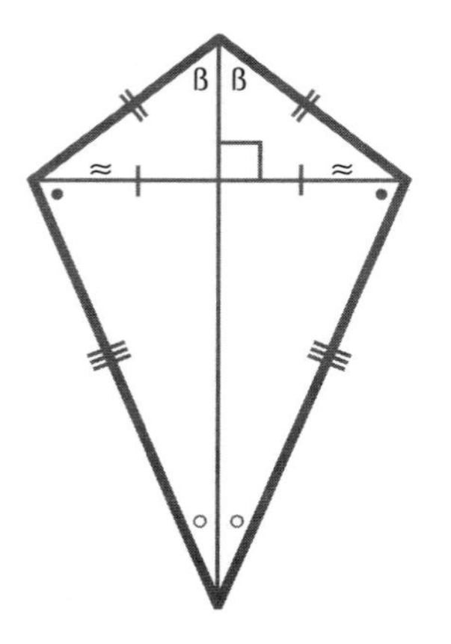

Kite

- angles marked with the same symbol are equal
- lines marked with the same symbol are equal
- the diagonals meet at a 90° angle
- one diagonal is bisected by the other

REMEMBER To bisect means to cut into two equal pieces.

The triangle

Special types of triangle

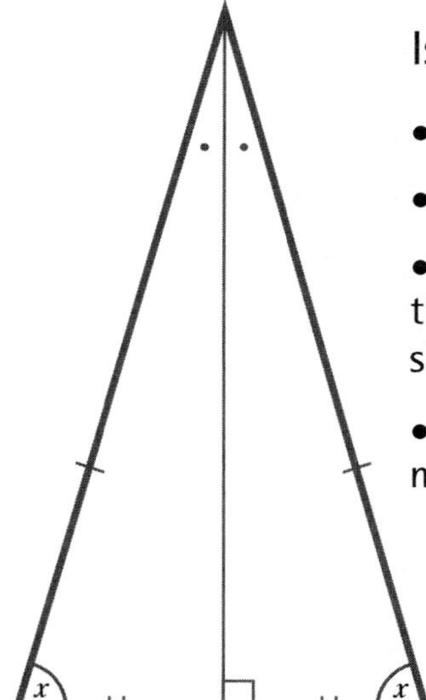

Isosceles triangle

- two sides are equal
- two angles are equal
- the line which bisects (cuts in half) the third angle will bisect the opposite side at 90°
- the two angles marked x are equal

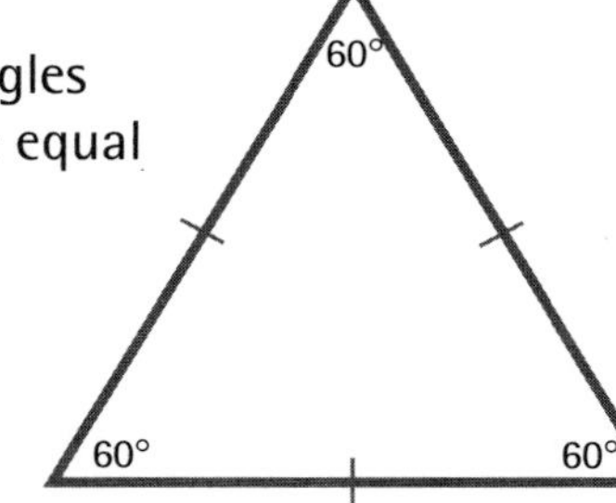

Equilateral triangle

- all the sides are equal
- all the angles are 60°
- a line which bisects any angle will bisect the opposite side at 90°

Finding the area of a triangle

$= \frac{1}{2}$ base x height

$= \frac{1}{2}\ bh$

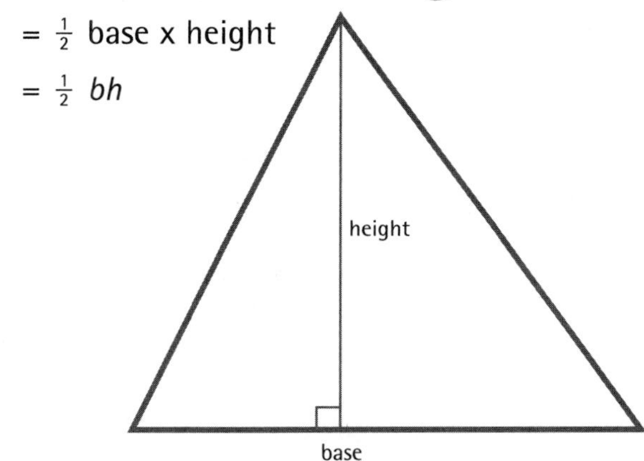

Practice questions

1) Calculate the area of this triangle.

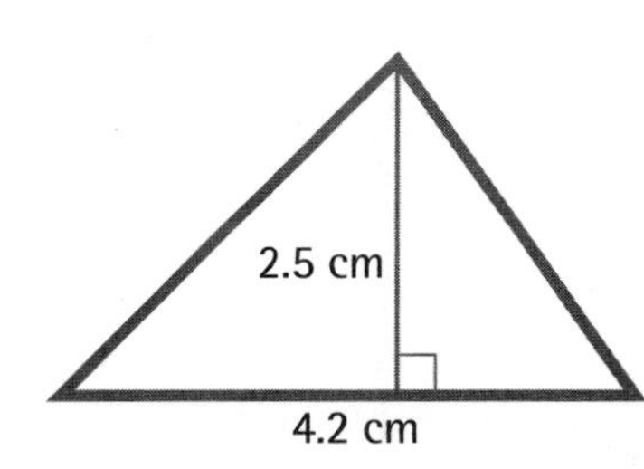

2) Calculate the sizes of all of the angles in this kite (which has not been drawn to the correct size).

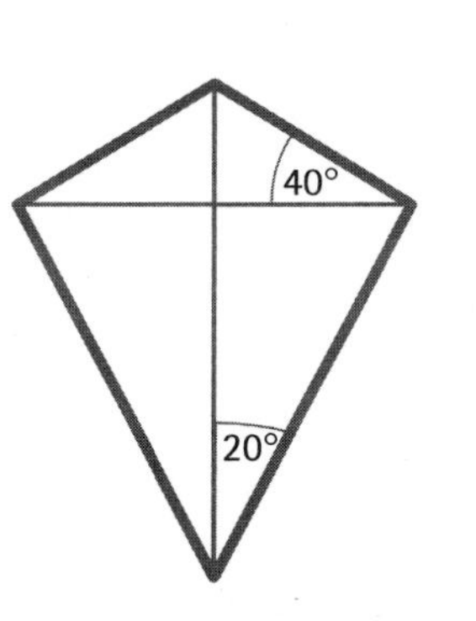

The circle

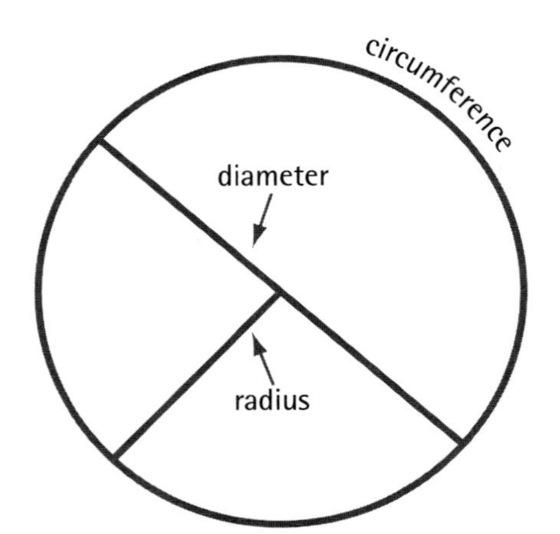

Circumference: the distance around the outside of a circle

Diameter: a line from one side of the circle to the other passing through the centre

Radius: a line from the centre of the circle to the circumference, which is half of the diameter

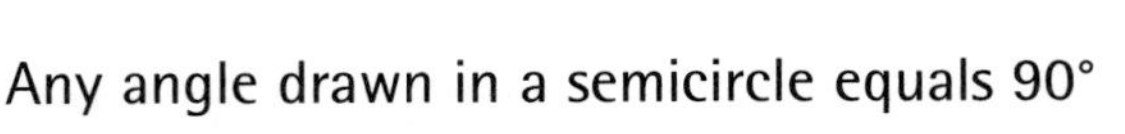

Any angle drawn in a semicircle equals 90°

A tangent is a straight line that just touches the circumference of the circle without cutting it.

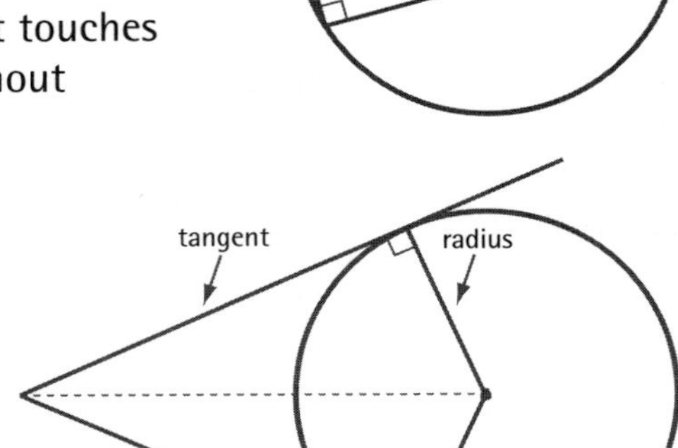

A radius always meets a tangent at 90°.

If you have two tangents from the same point then the line from the point to the centre of the circle is called an 'axis of symmetry'. (In this drawing, the line is shown by a dotted line.)

Finding the circumference

> **REMEMBER** Find the π key on your calculator. Use it in your calculations.

The formula list at the front of the **General Level** exam paper gives you:

Circumference of a circle: $C = \pi d$

Where: C = circumference

d = diameter

What does this mean?

It means that you have to multiply the diameter of the circle by a special number called 'pi' – written as π. This number is 3.14 to 2 decimal places, but using your calculator will give a more accurate value.

What if you are only given the radius of the circle?

Just remember to multiply the radius by 2 and you have the length of the diameter.

Example

Calculate the circumference of a circle with a radius of 4.2 cm.

Step 1: Find the diameter of the circle.

diameter = 4.2 x 2 = 8.4 cm

Step 2: Use the formula to find the circumference.

$C = \pi d = \pi \times 8.4 = 26.39$ cm

(Using $\pi = 3.14$ would have given $C = 26.38$ cm)

Finding the area of a circle

The formula list at the front of the **General Level** exam paper gives you:

Area of a circle: $A = \pi r^2$

Where: A = area

r = radius

This means that to find the area of a circle you multiply π by the radius squared.

REMEMBER
r^2 means $r \times r$

Example

Find the area of a circle with a radius of 8 cm.

Use the formula:

Area = πr^2

Area = $\pi \times 8^2 = \pi \times 64 = 201.06$ cm^2

(Using 3.14 would have given you an area of 200.96, which is not as accurate.)

REMEMBER
cm^2 means square centimetres

Semicircles and quarter circles

It is more common at **General Level** to be asked to find the perimeter or area of part of a circle. To do this just work out the circumference or area of the whole circle and divide it by 2 for a semicircle or by 4 for a quarter circle.

REMEMBER
A semicircle is a half circle.

Practice question

1) Find the area of this breakfast bar, which is made from a rectangle and $\frac{3}{4}$ of a circle.

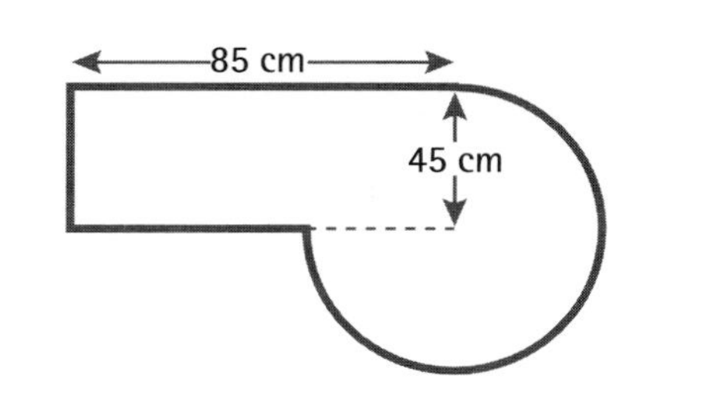

Calculating the area of shapes

Area of a square
= length x length
= l^2

l

l

Area of a rectangle =
length x breadth = lb

b

l

Area is the measure of a flat space or a 2-D shape. 2-D means 2-dimensional, that is flat shapes like the diagrams in this book.

From these two basic areas and the area of the circle (page 34) and the area of a triangle (page 33) you can find the areas of more complicated shapes.

These are called **composite shapes**.

REMEMBER Area is in square units, so there has to be a 2 (squared) in the answer.

REMEMBER When questions are broken into parts, the first part is usually easy and helps you to find the answer to other parts.

REMEMBER Look for ways to split up the shape into areas that you can work out, as they did with the fish tank in the TV programme.

Example 1

The diagram shows the floor plan of an office, marked **A** on the diagram, and the storage room joined to it, marked **B**.

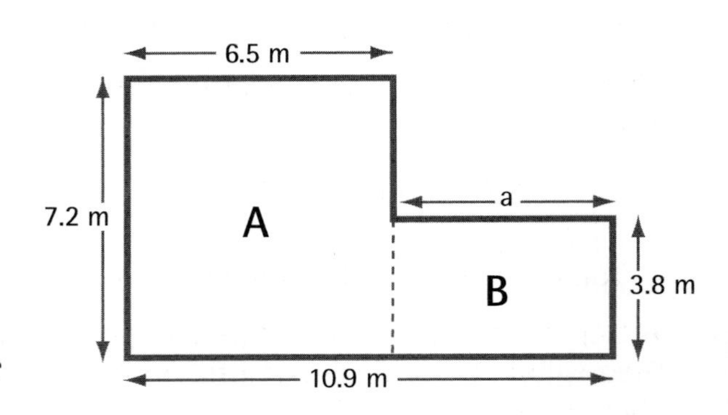

a) Find the area of carpet needed to cover the office floor.

b) Find the area of carpet needed to cover the storage room floor.

c) So, find the total area of carpet needed to cover both floors.

Step 1: Find the area of A.

Area of A = 7.2 x 6.5 = 46.8 m^2

Step 2: Find the area of B.

Before you can find the area of this shape, you must find out the length marked a:

a = 10.9 - 6.5 = 4.4 m

Area of B = 4.4 x 3.8 = 16.72 m^2

Step 3: Calculate the total area.

Total area = 46.8 + 16.72 = 63.52 m^2

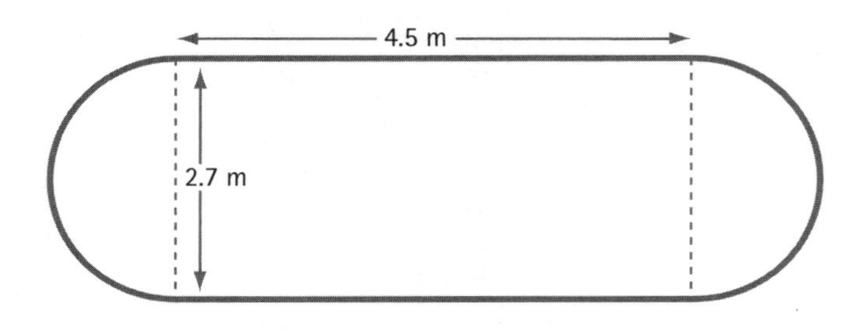

Example 2

Find the area of this shape. It is made from 2 semicircles and a rectangle.

Step 1: Split the shape into areas you can work out.

The two semicircles can be thought of as just one circle with a diameter of 2.7 m.

Step 2: Work out the area of the rectangle.

The rectangle is 2.7 m by 4.5 m.

Area of the rectangle = 2.7 x 4.5 = 12.15 m^2

Step 3: Work out the area of the two semicircles (equal to one circle).

Radius of circle = 2.7 ÷ 2 = 1.35 m

Area of circle = π x 1.35^2 = 5.73 m^2

Step 4: Work out the total area.

Area of the rectangle + area of the two semicircles = 12.15 + 5.73 = 17.88 m^2

Area of 2-D shapes

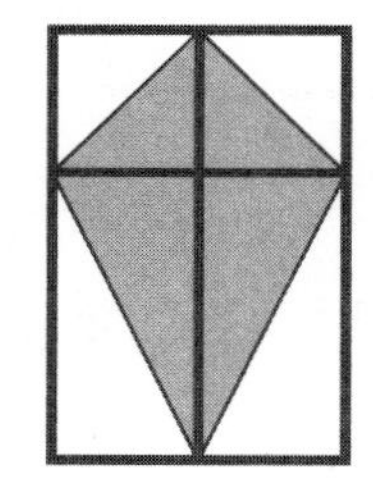

- A kite has an area of $\frac{1}{2}$ the length x the width.

This is because the kite can be split up into four triangles, each of which are half of the rectangle drawn round them.

So the kite is half of the area of the large rectangle.

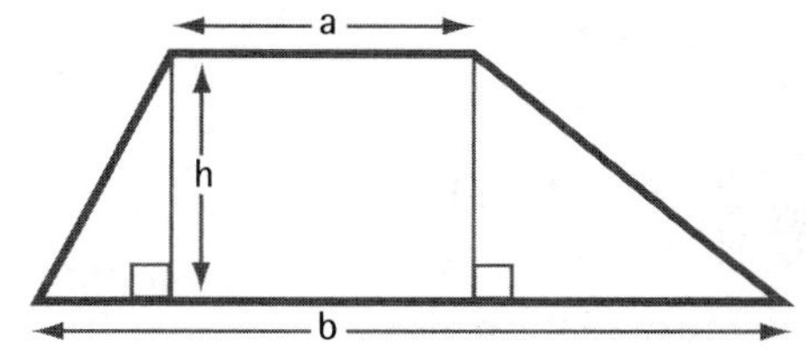

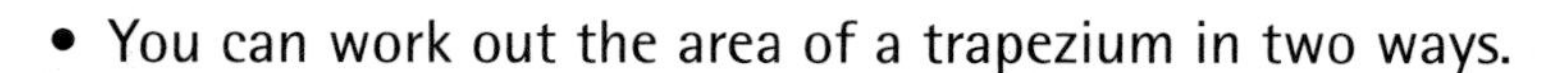

- You can work out the area of a trapezium in two ways.

A trapezium has one pair of parallel sides. You can work out the area of the two triangles and the rectangle in the centre and add them together.

Or you can remember the formula for working out the area of a trapezium:

$A = \frac{(a + b)}{2} \times h$ where A = area, h = height, a and b = the sides as above

This **Credit Level** formula makes finding the area of a trapezium easier.

Area of solid shapes

To work out the curved surface area of a cylinder the formula list at the front of the General Level exam paper gives you:

Curved surface area of a cylinder = $2\pi rh$

Where: r = radius h = height

This formula only lets you work out the curved part of the cylinder, not the top and bottom circles.

REMEMBER If you are given the diameter, remember to divide it by 2 to give the radius.

Example

Work out the total surface area of this cylindrical tin.

Area of base = $\pi r^2 = \pi \times 3.7^2$ = 43.01 cm^2

Area of top and bottom of tin = 2 x 43.01 = 86.02 cm^2

Curved area = $2\pi rh$ = 2 x π x 3.7 x 11.8 = 274.32 cm^2

Total surface area = 86.02 + 274.32 = 360.34 cm^2

Volume

Volume is a measure of the amount of space inside a 3-D object, such as a box or a can. Volume is measured in cubic units, i.e. cubic metres (m^3) and cubic centimetres (cm^3).

The volume of a cuboid = length x breadth x height

= Area of base x height

= $A \times h$

Volume of a cuboid = Ah

Prisms

A prism is a solid with a uniform cross-section. That means that wherever you slice it, every slice will have the same surface shape. (Uniform means the same.)

Cylinders

A cylinder is a type of prism. Think back to the can of Jock's Baked Beans. If you cut slices along the whole length of the can, each slice would look the same, like circles.

The end of the cylinder (the base) is a circle.

The formula list at the front of the **General Level** exam paper gives you:

Volume of a cylinder: $V = \pi r^2 h$

Where: V = volume

r = radius

h = height

REMEMBER Always show your working out, even if you are using a calculator. **No working = no marks!**

Find the volume of the can of Jock's Baked Beans.

Volume = area of base x height

= area of circle (πr^2) x height

= $\pi \times 3.7 \times 3.7 \times 11.8$

= 507.5 cm^3

Triangular prisms

A triangular prism is like the box that surrounds a very famous chocolate bar.

With a triangular prism you can work out the volume by calculating the area of the triangular end and multiplying it by the length (or height).

The formula list at the front of the **General Level** exam paper gives you:

Volume of a triangular prism: $V = Ah$

Where: V = volume

A = area

h = height

Find the volume of this triangular prism.

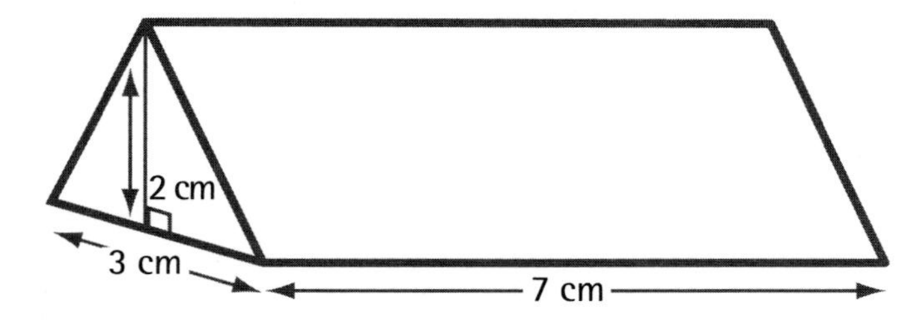

Area of triangle = $\frac{1}{2}bh = \frac{1}{2} \times 3 \times 2 = 3 \text{ cm}^2$

Volume of prism = $3 \times 7 = 21 \text{ cm}^3$

Example

In this example the side of a swimming pool is a composite shape.

Work out its volume in litres of water.

The dotted line has been drawn in to help you.

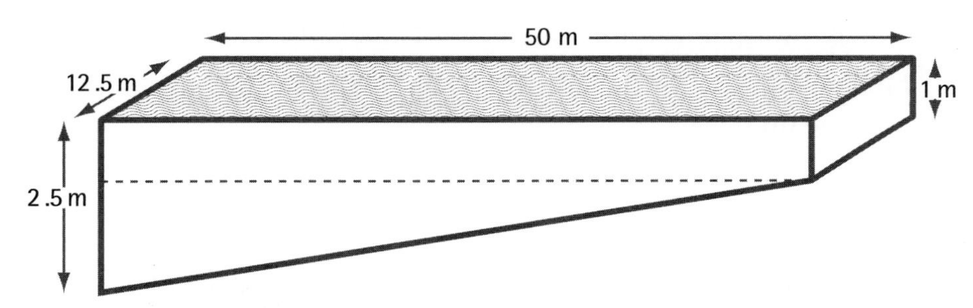

Height of the triangle = $2.5 - 1 = 1.5 \text{ m} = 150 \text{ cm}$

Area of the triangle = $\frac{1}{2} \times 5000 \times 150 = 375\,000 \text{ cm}^2$

Area of the rectangle = $5000 \times 100 = 500\,000 \text{ cm}^2$

Total area $= 375\,000 + 500\,000 = 875\,000 \text{ cm}^2$

Volume $= 875\,000 \times 1250$

$= 1\,093\,750\,000 \text{ cm}^3$

$= 1\,093\,750\,000 \div 1000$

$= 1\,093\,750$ litres

REMEMBER If you change all the measurements into centimetres at the beginning, it makes the answer easier to change to litres: 50 m = 5000 cm, 1 m = 100 cm

REMEMBER Check carefully what you put into your calculator when you work with lots of zeros.

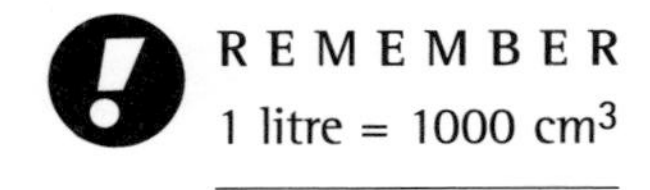

Coordinates and symmetry

Coordinates

In this diagram, A has coordinates (6, 3), B is (-3, 5), C is (-4, -3) and D is (0, -2).

Remember that the first coordinate goes across, then the second coordinate goes up or down

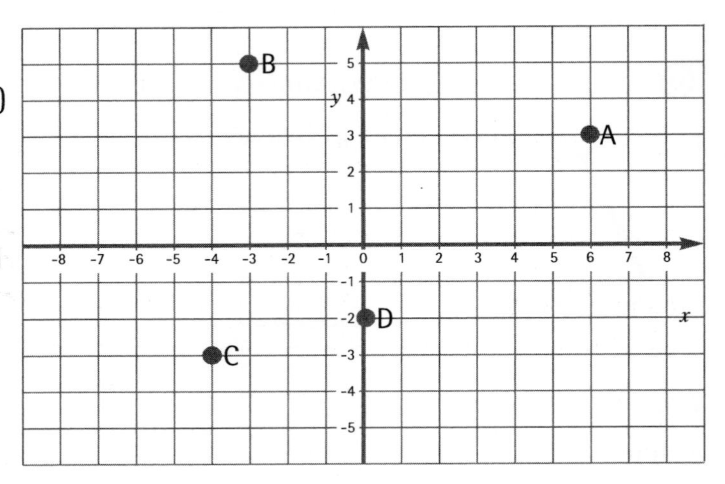

REMEMBER 'Along the hall and then up or down the stairs' will help you to remember how to plot coordinates.

Symmetry

Reflection symmetry

This is created when a design or picture is reflected in one or two lines. The reflection line is always drawn for you, here it is the dotted line.

Imagine standing a mirror on the dotted line and drawing the reflection of the design on the other side.

REMEMBER Use a ruler and be very accurate. Count your squares carefully.

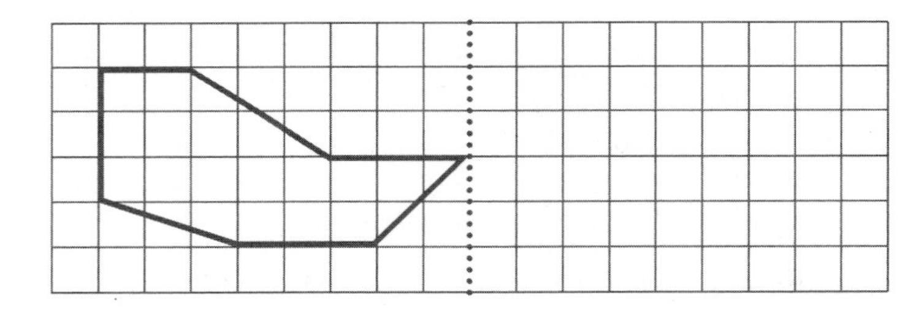

This is the finished drawing.

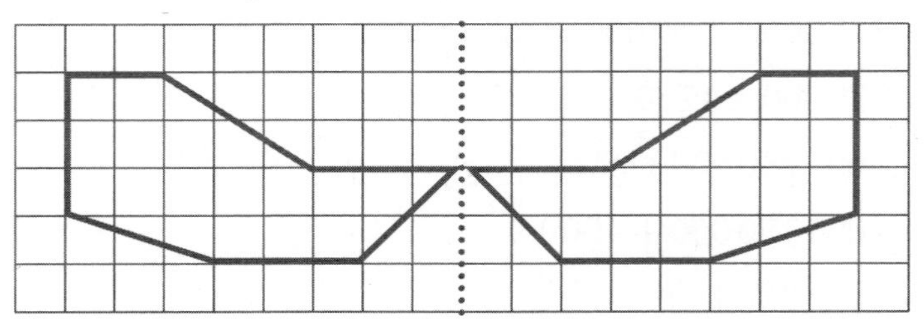

Notice that the shape is exactly the same. It has only been reflected in the line of symmetry.

Rotation symmetry

This is created by a shape turning on a single point.

The following design has to have half turn symmetry about the dot.

This can also be called a rotation through 180°.

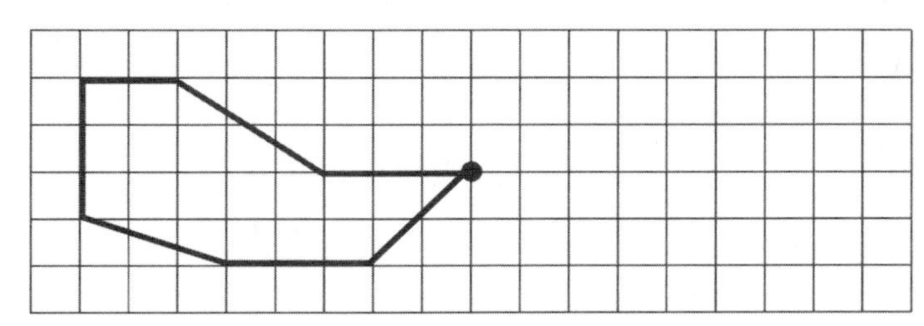

Imagine that the dot is a pin attaching the design to the grid. The design has to be turned through a half turn, i.e. a 180° rotation.

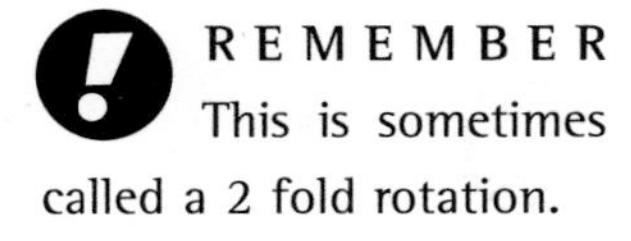

REMEMBER This is sometimes called a 2 fold rotation.

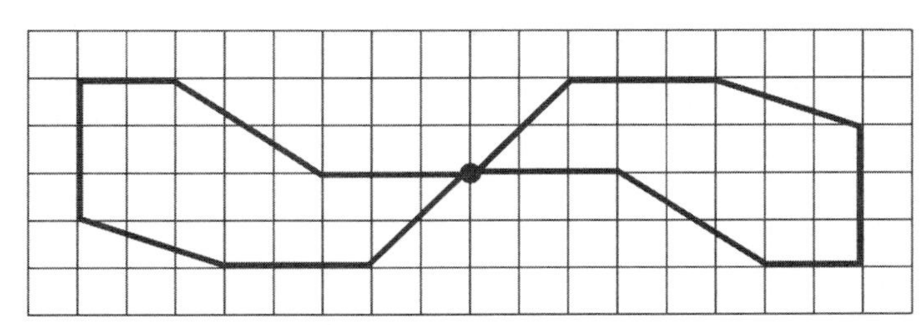

This is what it looks like after the rotation has taken place. Be careful to make the rotated shape exactly the same as the original shape.

Quarter turn symmetry means four turns within 360°.

In other words, the shape rotates through 90° each time.

There will be four identical rotated designs in the complete drawing.

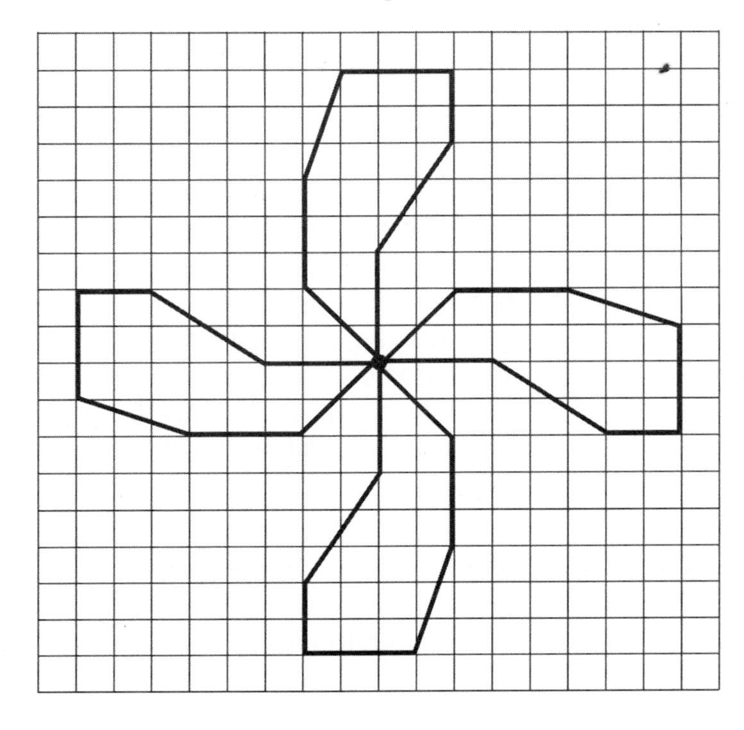

REMEMBER This is sometimes called a 4 fold rotation.

REMEMBER A 3 fold rotation would need 3 turns through 360°.

Right angled triangles

This section is about:

- using Pythagoras' theorem to find the length of a side in a right angled triangle
- using trigonometry to find the length of a side
- using trigonometry to find the the size of an angle
- solving problems with trigonometry

At the beginning of the **General Level** exam paper, you are given a formulae list. The formulae that are covered in this unit are shown in the FactZONE opposite.

Although you don't have to memorise these, it is much easier if you can. This section will show you how to use Pythagoras' theorem and trigonometry. These subjects sometimes cause students difficulties, so make some extra time to work carefully through this section. First of all, you need to understand the ideas. Then you need to practise using the ideas in the exam-style questions at the end of the book.

Return to this section and work through it again just before the exam. Ask your teacher for extra practice questions if you feel they would help.

You need to know how to calculate the length of the sides and the size of the angles in a right angled triangle. The triangle must have a right angle in it. If there doesn't appear to be a right angle, check the question carefully – it might be an angle in a semi-circle or it might not be a Pythagoras or trigonometry question at all.

When you are using Pythagoras' theorem, remember that squaring means multiplying by itself, so 3^2 means 3 x 3 not 3 x 2.

Again, you need to know how to round answers to a number of decimal places or to a significant figure. You also need to practise rounding to a sensible degree of accuracy. Ask your teacher if you are unsure about rounding. Be careful not to round too early in your working out. Leave answers to the earlier steps of your working in your calculator. Continue to use the full number, rather than pressing clear and keying in the rounded number.

Remember, this section is all about right angled triangles. The maths in this section can only be used with right angled triangles, not just any old triangle!

Pythagoras' theorem

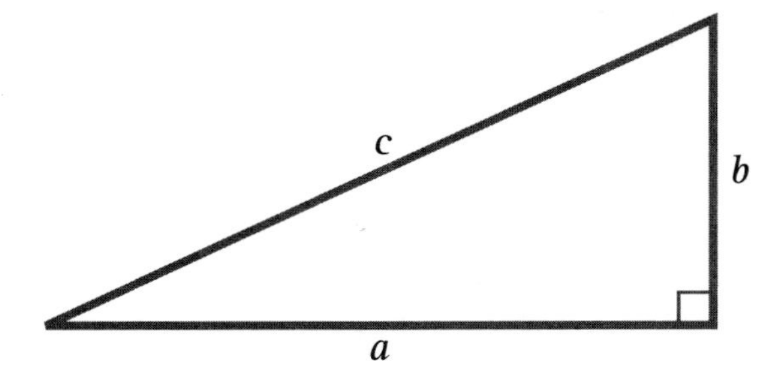

$a^2 + b^2 = c^2$

Trigonometric ratios in right angled triangles

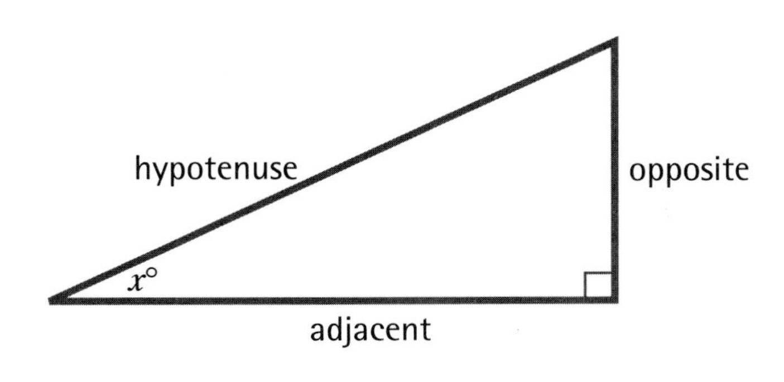

$$\tan x° = \frac{\text{opposite}}{\text{adjacent}}$$

$$\sin x° = \frac{\text{opposite}}{\text{hypotenuse}}$$

$$\cos x° = \frac{\text{adjacent}}{\text{hypotenuse}}$$

The 'trig word'

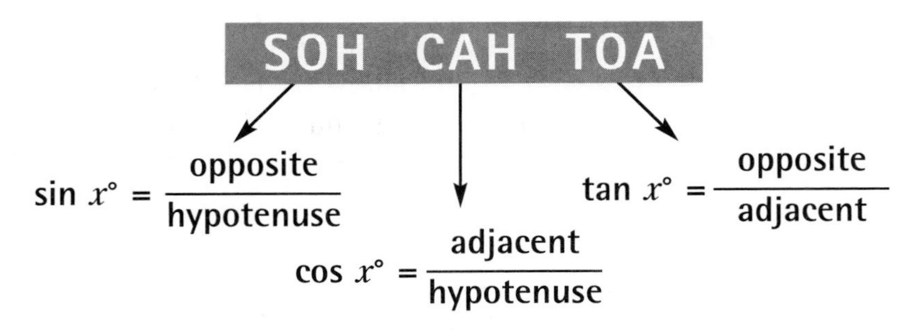

Pythagoras' theorem

> **REMEMBER** Imagine you are an archer standing in the right angled corner of a triangle. You shoot an arrow. Which side does it hit? The hypotenuse!

> **REMEMBER** It doesn't matter which way round a right angled triangle is drawn, the hypotenuse is always the side opposite the right angle and the longest side in the triangle.

Pythagoras was a Greek who lived about 500 BC. He developed a theory about right angled triangles.

Pythagoras' theorem says that the square of the hypotenuse is equal to the sum of the squares of the other two sides.

What is the hypotenuse?

It is the longest side of a right-angled triangle. It is always opposite the right angle.

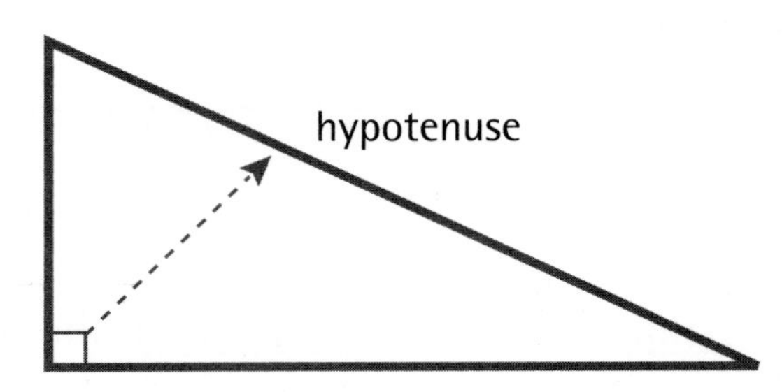

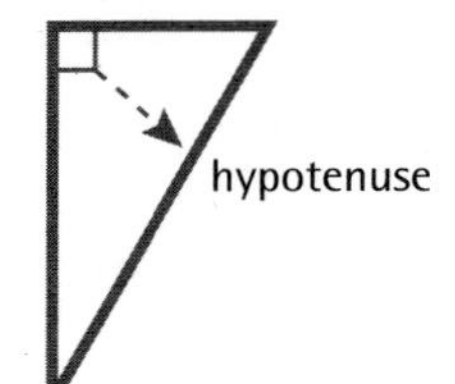

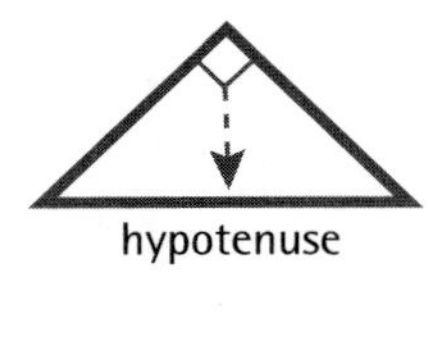

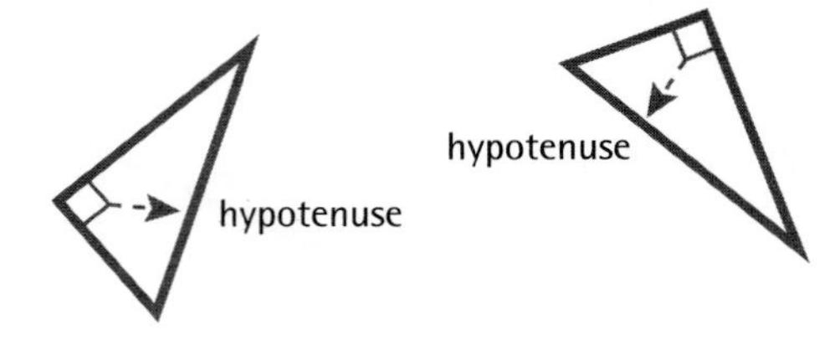

What does Pythagoras' theorem mean?

It means that when you:

square the lengths of the two shorter sides ($4 \times 4 = 16$ cm^2; $3 \times 3 = 9$ cm^2) and then add them together ($16 + 9 = 25$ cm^2), the answer is equal to the length of the hypotenuse squared ($5 \times 5 = 25$ cm^2).

This is easier to do than it sounds!

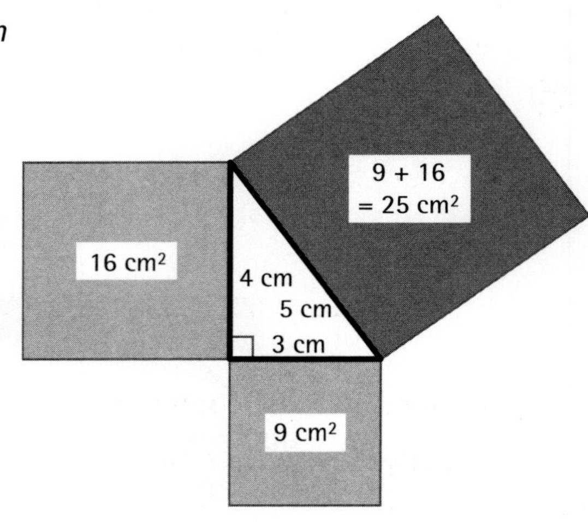

Finding the length of the hypotenuse

Example 1

What is the length marked x cm?

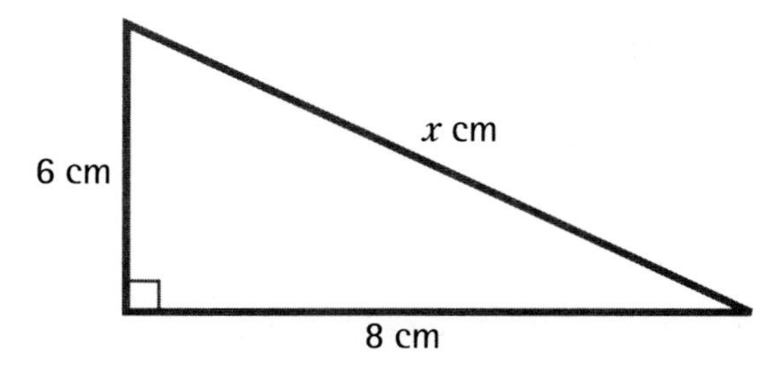

Step 1: Identify the hypotenuse.

It is the side opposite the right angle, so it is the side marked x cm.

Step 2: Write down Pythagoras' theorem.

It is on the exam paper if you can't remember it.

$a^2 + b^2 = c^2$

Step 3: Use Pythagoras' theorem with the numbers you have been given.

$a = 6$ cm and $b = 8$ cm

$6^2 + 8^2 = x^2$

$36 + 64 = x^2$

so $100 = x^2$

Is this the answer to the question?

You don't want x^2, you want x. To do this you need to take the square root of both sides:

$x = \sqrt{100}$

so $x = 10$ cm

REMEMBER $3 \times 3 = 9$ so $\sqrt{9}$ (read as the square root of 9) = 3

Example 2

A ramp is being designed for a wheelchair user. The height and length of the ramp are shown on the diagram below.

What is the length of the slope?

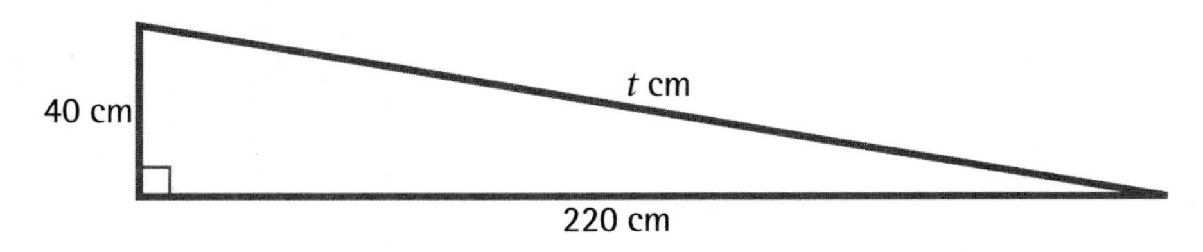

Step 1: Identify the hypotenuse.

The hypotenuse is the side marked t cm.

Step 2: Use Pythagoras' theorem to find the missing side.

$a^2 + b^2 = c^2$

$40^2 + 220^2 = t^2$

$1600 + 48\,400 = t^2$

$50\,000 = t^2$

$t = \sqrt{50\,000}$

$t = 223.6$ cm

The length of the slope is 223.6 cm or 2.236 m

REMEMBER On your calculator you will have 223.607 You then have to round it to one decimal place.

Finding the length of a shorter side

Pythagoras' theorem can also be used to find the length of a side that is not the hypotenuse.

What is the difference between finding a shorter side and finding the hypotenuse?

To find the hypotenuse you squared the two short sides and added them together.

To find one of the shorter sides you square the hypotenuse and subtract the square of the other short side from it.

In other words, you rearrange the theorem to give:

$a^2 = c^2 - b^2$ or $b^2 = c^2 - a^2$

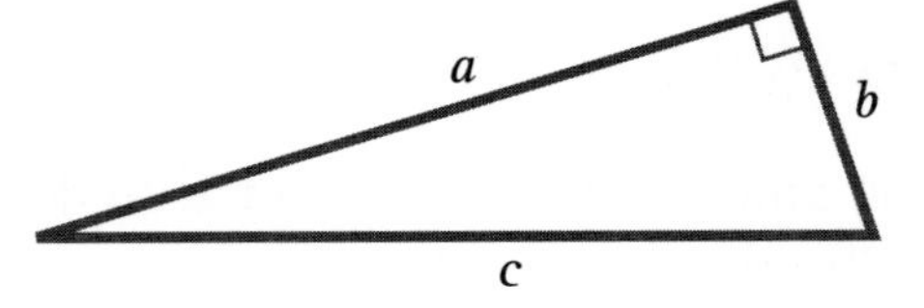

Example 1

To find the length of the short side y

you rearrange Pythagoras' theorem ($a^2 + b^2 = c^2$)

to give $a^2 = c^2 - b^2$

So using the numbers you are given in this problem:

$y^2 + 12^2 = 13^2$ becomes $y^2 = 13^2 - 12^2$

$y^2 = 169 - 144$

$y^2 = 25$

$y = \sqrt{25}$

$y = 5$

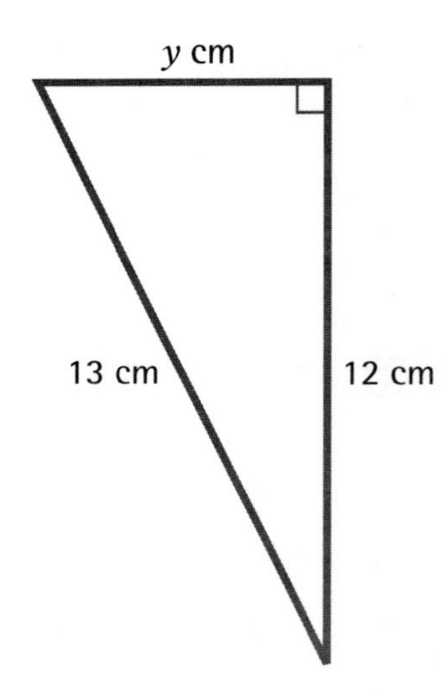

REMEMBER You have to find the square root in order to find the answer.

Example 2

In some triangles, the length of the sides are not easy-to-use whole numbers, instead they involve decimals. The same rule still works. Look at this triangle.

$b^2 = 8.3^2 - 2.5^2$

$b^2 = 68.89 - 6.25$

$b^2 = 62.64$

$b = \sqrt{62.64}$

$b = 7.9$ m

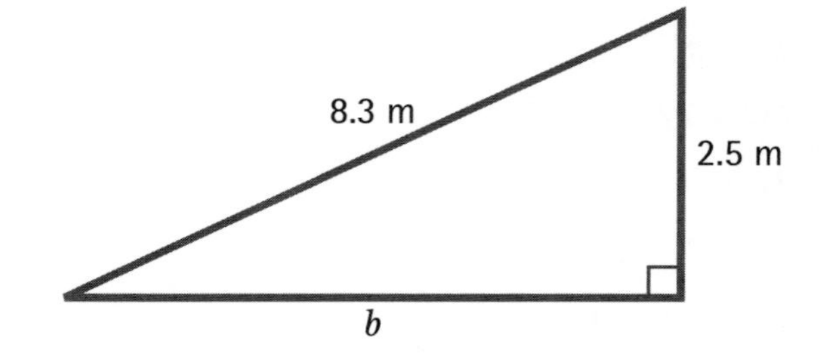

Using Pythagoras' theorem to solve problems

This is a Reasoning and Applications question. At first glance, is does not appear to be a Pythagoras problem, but it is.

There is a jigsaw measuring 26 by 21 inches.

The table has a diameter of 32 inches.

Will the completed jigsaw fit on the table?

Step 1: Redraw the diagram.

Drawing a diagram of the shape you are dealing with makes the question easier to understand.

The diagonal is the longest length in the jigsaw. If the diagonal of the jigsaw is 32 inches or less then it will fit onto the table.

If you draw a diagonal line on your rectangle then you have two right angled triangles. The diagonal, d, is the hypotenuse.

REMEMBER Sometimes you have to look hard at a question to see which maths 'tool' you need to use.

Step 2: Use Pythagoras' theorem.

$d^2 = 21^2 + 26^2$

$d^2 = 441 + 676$

$d^2 = 1117$

$d = \sqrt{1117}$

$d = 33.4$ inches

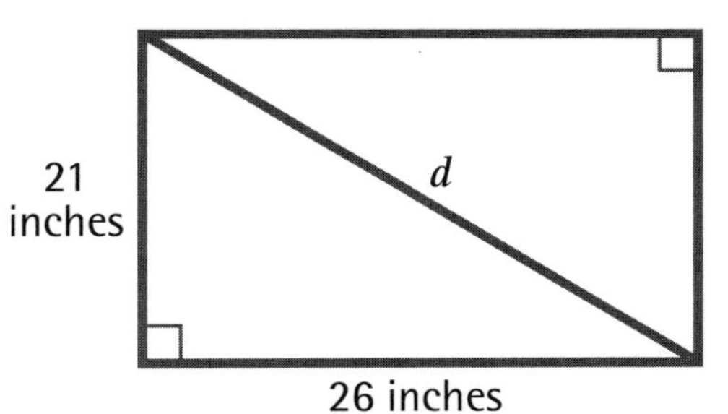

Step 3: Use your working out to answer the question.

The diagonal length of the jigsaw is 33.4 inches, the table is 32 inches, so the jigsaw is too big for the table by about 1.4 inches.

However, not all triangles have whole numbers to calculate with:

Calculate the height of the mast, marked h, on this yacht.

The hypotenuse is 3.4 m. Use Pythagoras' theorem to find out the length of the short side h.

$h^2 = 3.4^2 - 1.6^2$

$h^2 = 11.56 - 2.56$

$h^2 = 9$

$h = \sqrt{9}$

$h = 3$ m

So the mast is 3 metres high.

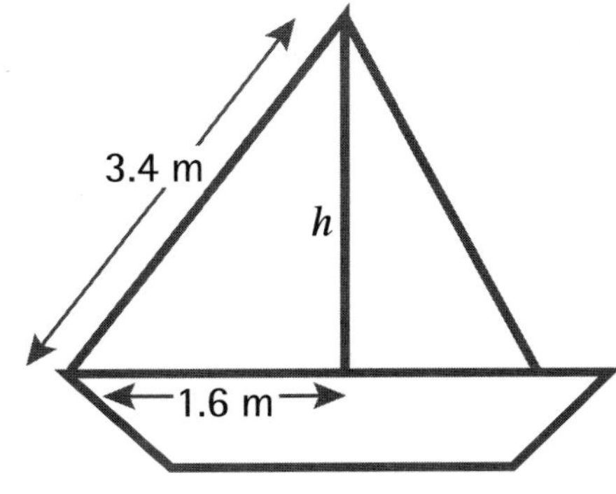

Right angled triangles

Trigonometry

Can you speak trig?

There is a great word to help you remember what to do in trigonometry.

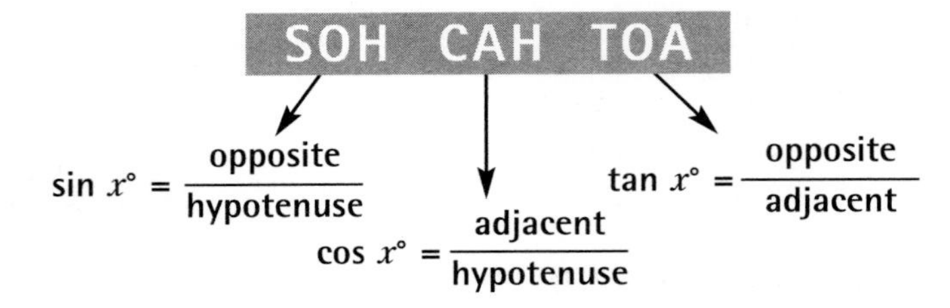

These examples will show you how to use SOH CAH TOA to help you solve problems in trigonometry.

First, think back to the basics. You know how to find the hypotenuse in a right angled triangle. It is the longest side and is opposite the right angle.

How do you find the **opposite** *and* **adjacent** *sides in a right angled triangle?*

There will always be a marked angle (other than the right angle) in a trig question. This will be either an angle whose size you know or an angle you have to work out.

In this drawing it is the angle marked $x°$.

REMEMBER Think back to the archer. If you shoot an arrow this time, you will hit the opposite side.

The **opposite** side is the side opposite the angle.

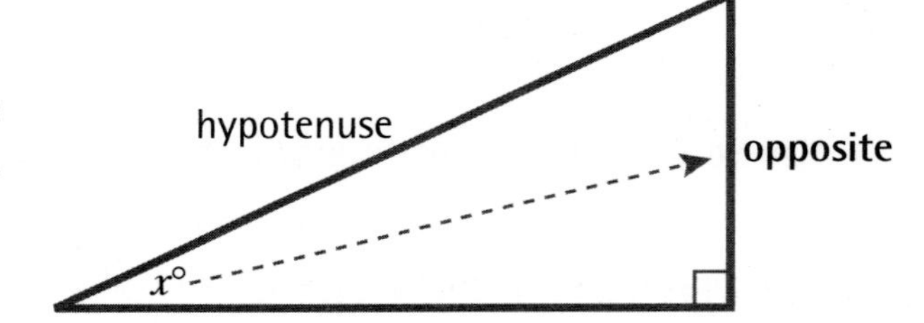

REMEMBER If you put your finger on the angle and move your finger along to the right angle, you are tracing along the adjacent side.

The **adjacent** side is the side next to the angle. (Adjacent means 'next to'.)

hypotenuse

opposite

$x°$

adjacent

Finding the length of a side of a triangle

Example 1

Find the length of the side marked x cm in the triangle.

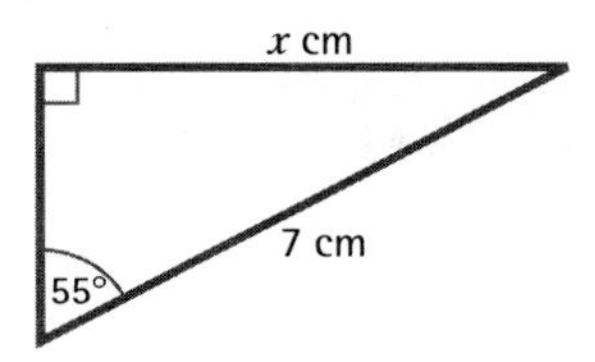

Step 1: Label the sides of the triangle you are given, so you know what you have to work with.

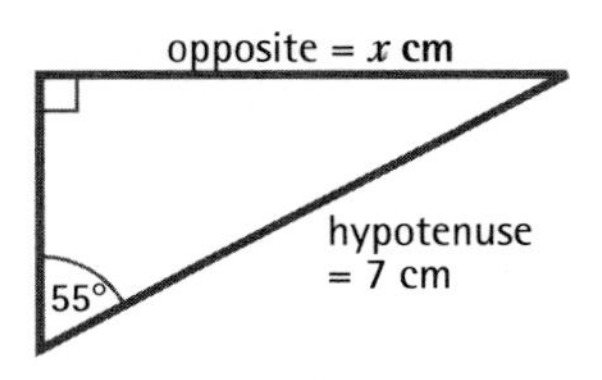

Step 2: Write down the 'trig word' SOH CAH TOA.

Tick off the sides you have to work with. Then circle the part with two ticks.

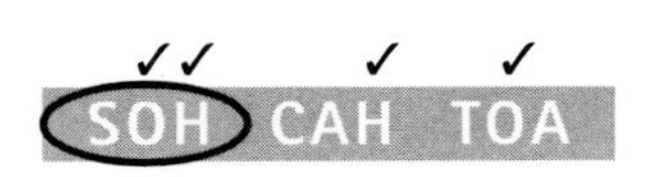

SOH has 2 ticks, so you are going to use sin.

Step 3: Do the working out.

$$\sin 55° = \frac{\textbf{opposite}}{\textbf{hypotenuse}}$$

$$\sin 55° = \frac{x}{7}$$

$$7 \times \sin 55° = x$$

$$x = 5.73 \text{ cm}$$

REMEMBER Check that your calculator is in degree mode – with DEG on the display.

Example 2

Find the length of the side marked y cm in the triangle.

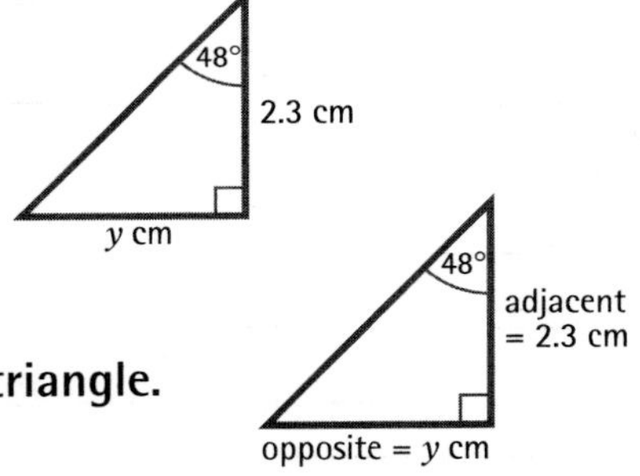

Step 1: Label the sides of the triangle.

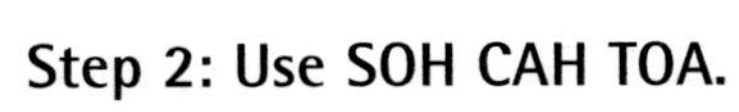

Step 2: Use SOH CAH TOA.

✓ ✓ ✓✓
SOH CAH TOA

TOA has two ticks, so you are going to use tan.

Step 3: Do the working out.

$$\tan 48° = \frac{\textbf{opposite}}{\textbf{adjacent}}$$

$$\tan 48° = \frac{y}{2.3}$$

$$2.3 \times \tan 48° = y$$

$$y = 2.55 \text{ cm}$$

Finding the size of an angle using trigonometry

Example 1

Find the size of the angle marked $a°$.

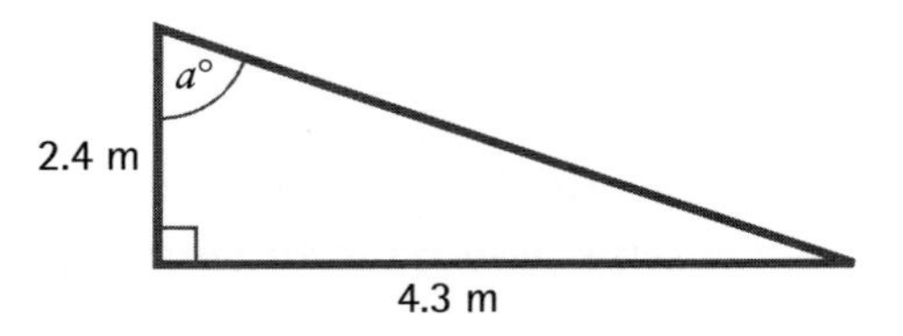

Step 1: Label the sides of the triangle you are given so that you know what to work with.

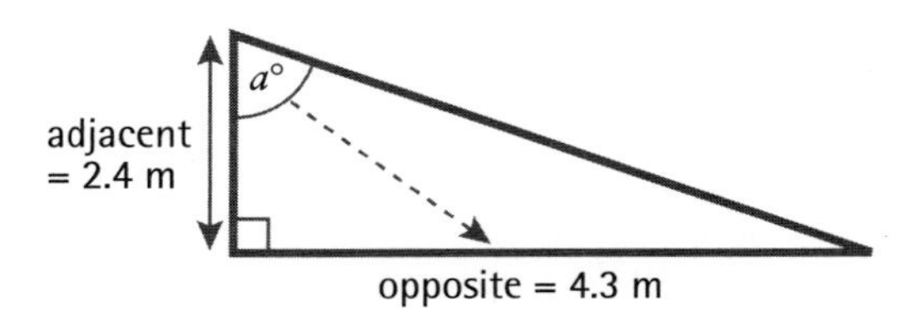

Step 2: Write down the 'trig word' SOH CAH TOA.

Tick off the sides you have to work with.
Then circle the part with two ticks.

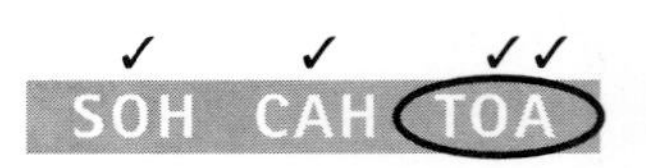

TOA has two ticks, so you are going to use tan.

Step 3: Do the working out.

$$\tan a° = \frac{\textbf{opposite}}{\textbf{adjacent}}$$

$$\tan a° = \frac{4.3}{2.4}$$

$\tan a°$ = 1.79 (to 2 decimal places – but leave the full value in your calculator)

REMEMBER If you are not sure how to do this on your calculator, remember to ask your teacher.

You will now need to use the SHIFT, INV or 2nd F key before the tan key to find the size of the angle, so:

$$a = 60.8°$$

Now try these examples yourself.

The answers are in the boxes below. Cover them up with a piece of paper until you have worked out your answers.

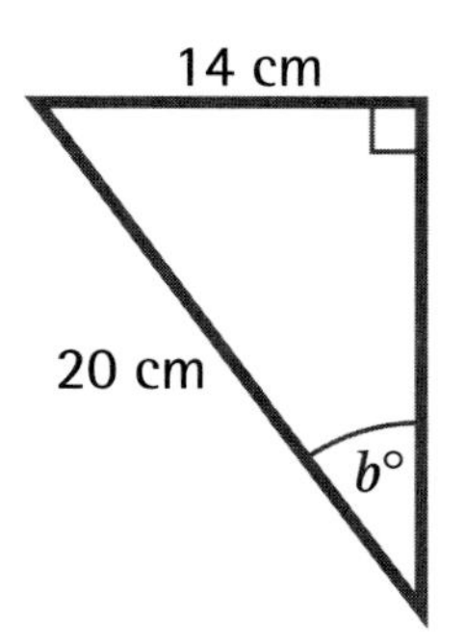

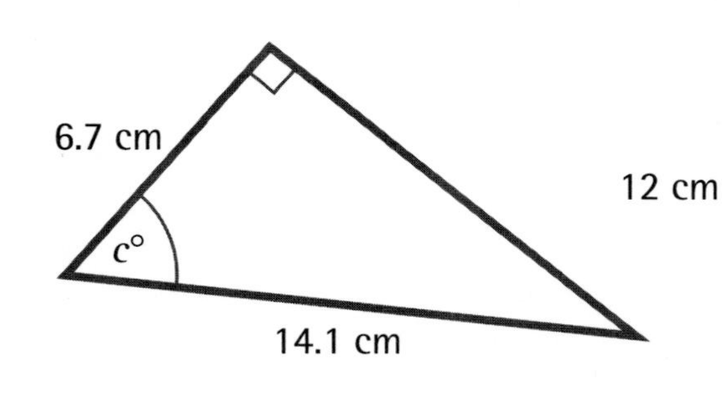

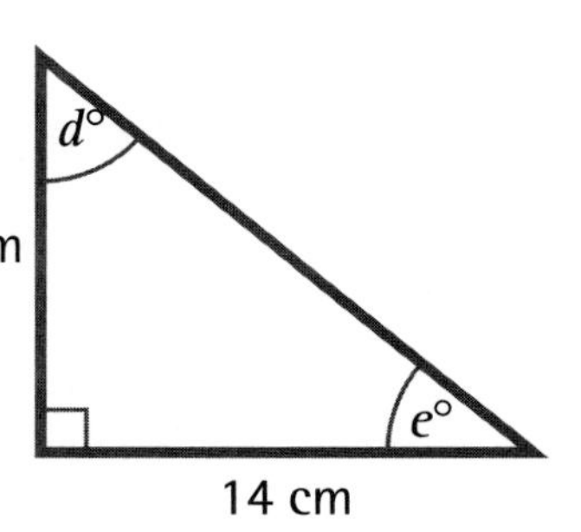

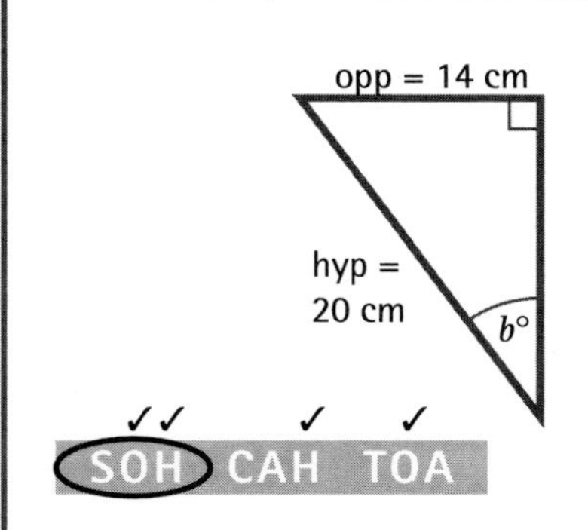

$$\sin b° = \frac{\text{opposite}}{\text{hypotenuse}}$$

$$\sin b° = \frac{14}{20}$$

$$\sin b° = 0.7$$

$$b = 44.4°$$

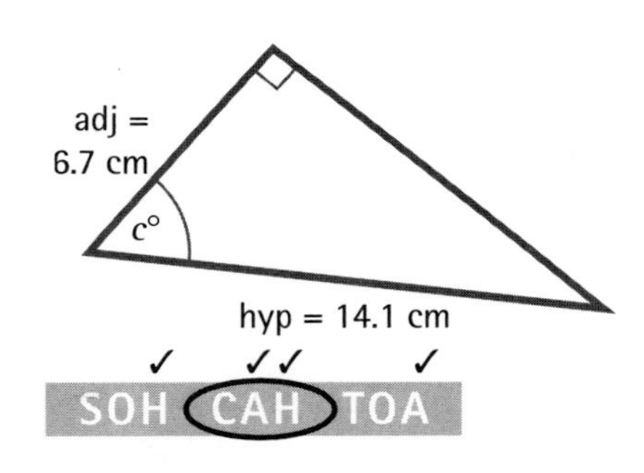

$$\cos c° = \frac{\text{adjacent}}{\text{hypotenuse}}$$

$$\cos c° = \frac{6.7}{14.1}$$

$$\cos c° = 0.48$$

$$c = 61.6°$$

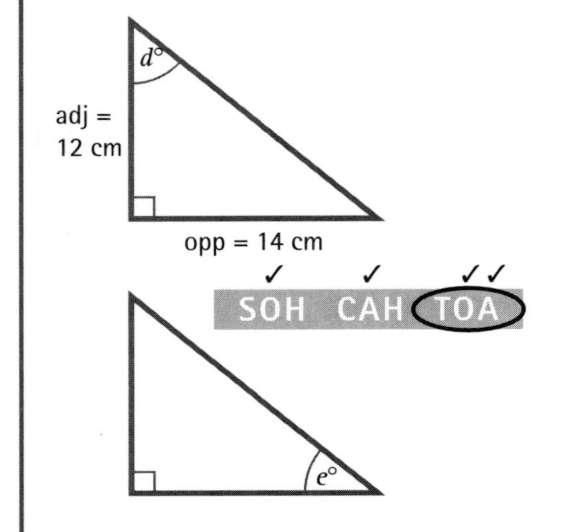

$$\tan d° = \frac{\text{opposite}}{\text{adjacent}}$$

$$\tan d° = \frac{14}{12}$$

$$\tan d° = 1.17$$

$$d = 49.4°$$

$$e = (180 - 90 - 49.4)°$$

$$e = 40.6°$$

REMEMBER Press INV or SHIFT or 2nd F before the sin, cos and tan keys to get the size of the angle.

REMEMBER You don't need trigonometry for angle $e°$. Even though there is a complicated way, simple methods still work!

Using trigonometry to solve problems

Example 1

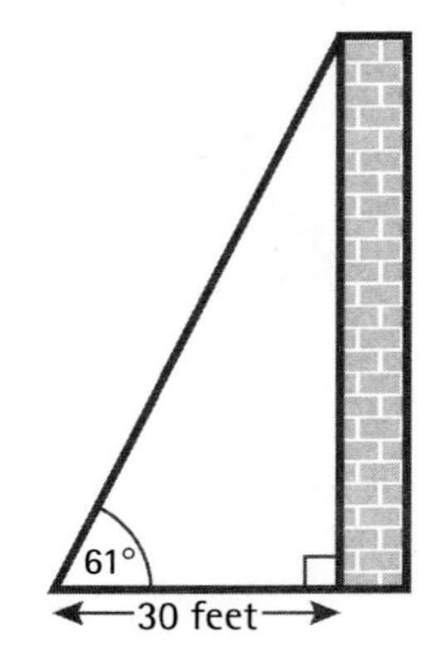

Scott stands 30 feet from the bottom of a tower.

He measures the angle of elevation from the ground to the top of a tower. It is 61°.

How high is the tower?

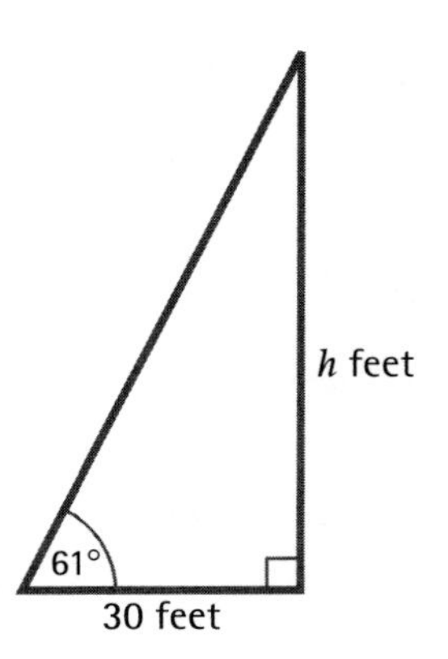

Step 1: Redraw the diagram and label the sides of the triangle you are given.

Step 2: Use SOH CAH TOA.

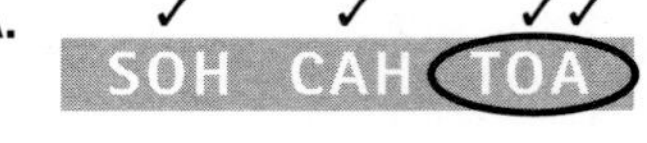

TOA has two ticks, so you are going to use tan.

Step 3: Do the working out.

$$\tan 61° = \frac{\textbf{opposite}}{\textbf{adjacent}}$$

$$\tan 61° = \frac{h}{30}$$

$$30 \times \tan 61° = h$$

$$h = 54.12 \text{ feet}$$

REMEMBER You don't need the SHIFT, INV or 2nd F key to work out a side of a triangle.

Example 2

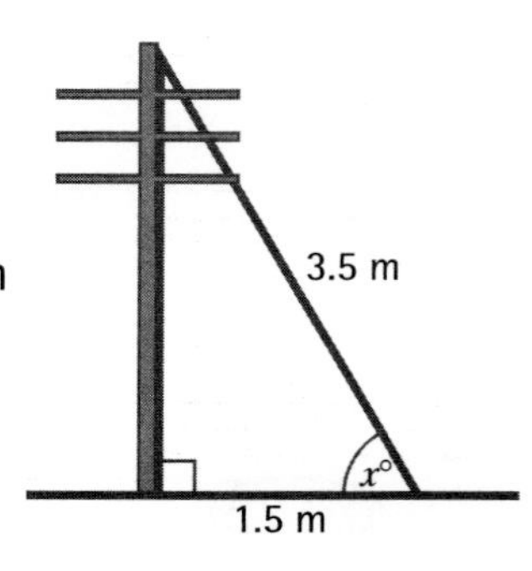

A cable 3.5 m long is used to secure a telegraph pole.

The cable is to be secured 1.5 m from the base of the pole.

Calculate the size of the angle marked x°.

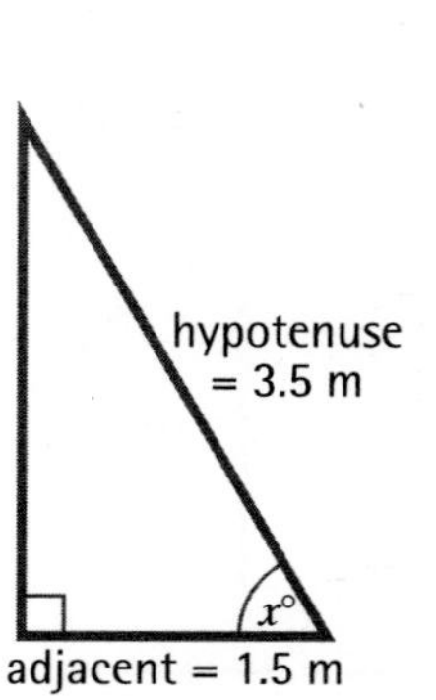

Step 1: Redraw the diagram and label the sides of the triangle you are given.

Step 2: Use SOH CAH TOA.

✓ ✓✓ ✓
SOH CAH TOA

CAH has two ticks, so you are going to use cos.

Step 3: Do the working out.

$$\cos x° = \frac{\text{adjacent}}{\text{hypotenuse}}$$

$$\cos x = \frac{1.5}{3.5}$$

$$\cos x° = 0.43$$

$$x = 64.6°$$

Example 3

A satellite installation company will only use a ladder to fit a satellite dish if the angle that the ladder makes with the ground is between 70° and 75°.

Will they be able to use a 4 metre long ladder to fit a dish 3.7 metres up a wall?

Step 1: There is no diagram, so you must draw one and label the sides of the triangle.

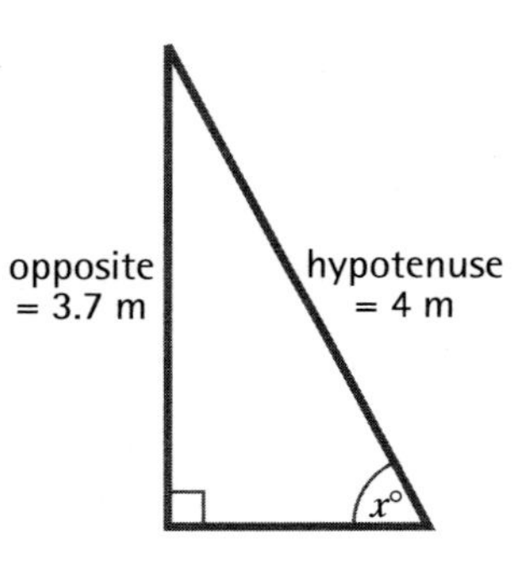

Step 2: Use SOH CAH TOA.

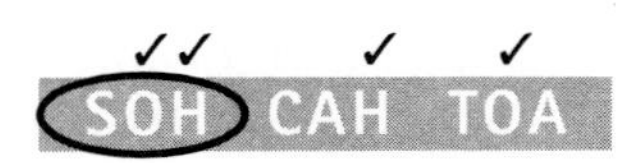

SOH has two ticks, so you are going to use sin.

Step 3: Do the working out.

$$\sin x° = \frac{\text{opposite}}{\text{hypotenuse}}$$

$$\sin x° = \frac{3.7}{4}$$

$$\sin x° = 0.925$$

$$x = 67.7°$$

Step 4: Answer the question in full.

No, the company cannot use the ladder because 67.7° is less than 70°.

REMEMBER Unsure if you should use Pythagoras or trigonometry? Remember that Pythagoras uses only sides, trigonometry needs an angle and sides.

Algebra

This section is about:

- using indices and scientific notation
- number patterns and how to describe them
- using formulae
- multiplying out brackets and simplifying
- factorising algebraic expressions
- solving equations
- solving inequalities

Indices and scientific notation are ways of working with very large or small numbers. An answer may appear on your calculator in scientific notation, so it is important to know what the display on your calculator means. You will certainly have a question in the exam on scientific notation and maybe one on indices as well, so make sure that you understand them.

Spotting patterns is important. It means that you can make predictions about what is going to happen. Being able to develop rules to describe patterns is an important skill. This section shows you how find the rule for any pattern.

Many people think algebra is hard and this fear creates a barrier in their mind which prevents them from understanding it. Algebra is simply a method of problem solving using letters instead of numbers. Algebra works with exactly the same rules as numbers do, so if you get stuck with algebra, you can often put in easy numbers instead and work out how to solve the problem.

There are several different skills that you will need in algebra. Formulas are rules that let you easily work out a numerical value, perhaps for a volume or a temperature. You will need to solve equations and inequalities, which are really just balancing puzzles. You will also have to understand how to multiply out expressions and factorise them, which has a lot to do with knowing your multiplication tables.

None of the work in this section is difficult, just keep an open mind and work through it. The emphasis here is on techniques, so study the examples in the section until you understand them. Then cover them up and try them for yourself. Remember to do the practice questions and the exam-style questions, too.

FactZONE

Algebra

$8a$ means $8 \times a$ — The multiplication sign is missed out.

5^2 means 5×5 — In just the same way: y^2 means $y \times y$

Variable

Variable is the name for the letter in algebra, for example m is the variable in $15m$.

Collecting like terms

In algebra you can only add or subtract terms that have the same variable, for example:

$3k + k + 10k = 14k$

$5x - 2x + 7y = 3x + 7y$

You cannot add or subtract terms that are different, so you can't add xs and ys together.

You cannot combine squared or cubed terms with ordinary terms, so $x + x^2$ cannot be added. The expression:

$9x^2 + 5x - 6x^2 + 7 + 6x$

can be rearranged to give:

$9x^2 - 6x^2 + 5x + 6x + 7$

$= 3x^2 + 11x + 7$

Notice that when you rearrange an expression, the plus or minus signs are 'stuck' to the term that follows them and so they move as a single unit.

Expression

Expression is the name given to letters and often numbers linked together, for example:

$2x + 7$ $6 - y$ $ab - b$ $7(2m - n)$

Equation

Equation is the name given to two expressions or an expression and a number linked with an equals sign, for example:

$6x + 5 = 23$ $2(a - 7) = 4a - 10$ $6 = 4c - 18$

Inequality

The same as an equation, but using an inequality sign instead of an equals sign. The inequality signs are:

$>$ greater than $<$ less than

$\geq$ greater than or equal to $\leq$ less than or equal to

Indices and scientific notation

Indices

Indices are the small numbers written above a number.

For example, the ^{3} in 4^3 shows that you multiply 4 by itself 3 times.

4^3 means 4 x 4 x 4 = 64

2^6 means 2 x 2 x 2 x 2 x 2 x 2 = 64. It is read as '2 to the power of 6'.

Example 1

Which is larger, 2^9 or 5^4?

2^9 = 2 x 2 x 2 x 2 x 2 x 2 x 2 x 2 x 2 = 512

5^4 = 5 x 5 x 5 x 5 = 625

So 5^4 is larger.

REMEMBER Your calculator is a powerful tool. Make sure you know how to use it.

You can work out indices with your calculator.

Look for the key with either x^y or y^x on it.

(It might be written above a key, this means you need to press the SHIFT or INV or 2nd F key first.)

For most calculators to work out 2^9 you would enter:

[2] [y^x] [9] [=]

Try this on your calculator and check that you get the correct answer of 512. If not, ask your teacher how to do this on your calculator.

Scientific notation

This is a way of writing down very large or very small numbers without a lot of zeros.

It is always written as:

$a \times 10^n$

where a and n are numbers.

On a calculator, it would be 5.3 E 7 or 5.3^{07}. This would be written as 5.3×10^7.

Here are some numbers written in scientific notation. Notice that the number before the decimal point is always a single digit between 1 and 10.

3.24×10^3 6.0×10^{-7} 1.578×10^4 8.7×10^{-2}

Example 1

Write 314 000 000 in scientific notation.

314 000 000 = 3.14 x 10 x 10 x 10 x 10 x 10 x 10 x 10 x 10 = 3.14×10^8

How is that done?

Write the number down with a decimal point after the first digit, but don't write down the long string of zeros. You now have 3.14

Now, imagine the original number had been written as 314 000 000.

Note the decimal point

Count how many spaces there are between where the decimal point was and where you have 'moved' it to:

3 . 1 4 0 0 0 0 0 0 .

It is 8 steps, so it is 10 to the power of 8, i.e. 10^8

Example 2

Write 0.000 043 in scientific notation.

As you are dividing by 10, the decimal point 'moves' 5 places to the right.

0. 0 0 0 0 4 3

If you move the decimal point from the left to the right, you must put a minus sign in front of the power.

So:

$0.000043 = 4.3 \div 10 \div 10 \div 10 \div 10 \div 10 = 4.3 \times 10^{-5}$

Example 3

Write 9.423×10^5 in full.

$9.423 \times 10^5 = 9.423 \times 10 \times 10 \times 10 \times 10 \times 10 = 942300$

Here you do the opposite of what you did in Example 1 above. Fill in the gaps with zeros.

REMEMBER You don't need to write in the decimal point when the answer is a whole number.

Practice questions

1) Write 5^6 as a whole number.
2) Write 0.000 054 5 in scientific notation.
3) Write 7.5 million in figures. Then write 7.5 million in scientific notation.
4) Write out the number 8.4×10^{-3} in full.

Patterns and how to describe them

Looking at patterns and describing them is a Reasoning and Applications skill. So you will need to think about the pattern, describe it mathematically and then work out the answers to any other questions you may be asked.

Example 1

Each dot represents a skater skating in formation.

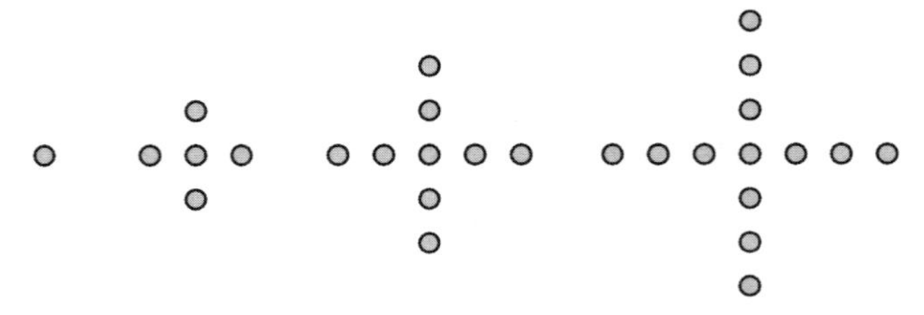

How can you find a rule?

Step 1: Count up how many dots there are in each pattern.

In this case, you get the number sequence: 1, 5, 9, 13

Step 2: Put these in a table and work out the difference between them.

Pattern number	Number	Difference
1	1	
		4 (5 - 1)
2	5	
		4 (9 - 5)
3	9	
		4 (13 - 9)
4	13	

This shows a constant difference (a difference that is always the same).

Step 3: Multiply the pattern number by the constant difference.

The first stage of the rule is to multiply the pattern number by the constant difference. In this case, the constant difference is 4.

Ask yourself, 'What do I need to add or subtract to make the number?' For pattern number 1: 1 x 4 = 4, so to make the 1 you need to subtract 3.

Check this works with the next pattern number: 2 x 4 = 8, then 8 - 3 = 5
This is the correct result, so it works and you have found the rule.

Step 4: Write out the rule.

Here it is: Number of dots = (Pattern number x 4) - 3

It is more usual to write the rule using algebra. The question might say:

- Write down the rule (or formula) for the number of dots, N, when you know the number of the pattern, P, in the sequence.

REMEMBER
A rule lets you work out any number in a pattern without working out all the previous numbers.

In this example:
Number of dots = (Pattern number x 4) - 3 can be written as $N = 4P - 3$

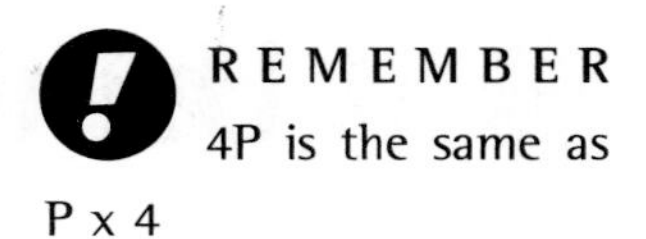

Example 2

Often the pattern is not shown as a set of pictures. In this example, there is a single drawing and a table to complete. The way to find the rule is the same.

REMEMBER It is important to show the equation in algebra. Show that you understand that $4P$ means 4 x P by writing it as $4P$

A small garden fence is made by joining long wooden posts with rope, as shown below.

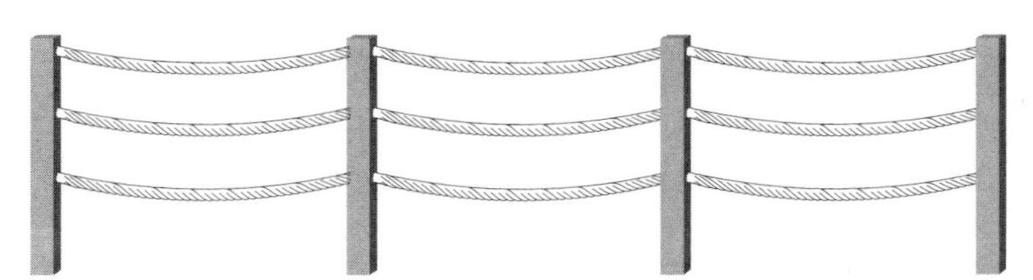

Copy and complete the following table.

Number of posts (P)	1	2	3	4		10
Number of pieces of rope (R)	0		6			

REMEMBER There are usually some questions like this in the exam. When you try past exam papers, practise making tables of difference and use the clues from the tables to find the rules.

The shaded box is there to remind you that part of the number sequence is missing. This is the completed table:

Number of posts (P)	1	2	3	4		10
Number of pieces of rope (R)	0	*3*	6	*9*		*27*

Write down a formula for the number of pieces of rope, R, when you know the number of posts, P.

The constant difference between the numbers in the first section of the table above is 3.

Multiply 1 (Post) by 3 and you get 3.

How do you get 0?

You take away 3.

Check with the next entry in the table:

2 x 3 = 6, 6 - 3 = 3, which is correct.

So the rule is $R = 3P - 3$.

If you have 63 ropes, how many posts would there be?

You will have to 'undo' the rule to work this out. To undo 'subtract 3', you need to add 3: 63 + 3 = 66

To undo 'multiply by 3', you divide by 3: 66 ÷ 3 = 22

So there are 22 posts.

REMEMBER If you need to remind yourself about solving equations, turn to page 64.

Using formulae

A formula is used to make calculations easier. You have already used formulae (the plural of formula) as you worked through this book.

Do you remember the formula list at the front of the General level exam paper? One of the formulae was used to calculate the volume of a cylinder:

$V = \pi r^2 h$

Each of the letters, or variables, stands for something:

- V is the volume
- r is the radius of the circular base
- h is the height of the cylinder
- π is the special number equal to approximately 3.14 that we looked at in the circle work in Section 2

If we know the values of r and h, we can work out the volume of the cylinder.

Example 1

Using the formula:
$V = \pi r^2 h$
find V if r = 3.4 and h = 5.2

Step 1: Substitute the values for r and h.

$= \pi \times 3.4^2 \times 5.2$

Step 2: Work out the value of V.

$= 188.85$

Example 2

In the TV programme archaeologists discovered a skeleton. They used a formula to find out approximately how tall the man was when he was alive by using the length of two leg bones, called the femur and tibia.

The formula for a white male was:

$h = 1.3\,(f + t) + 63.29$

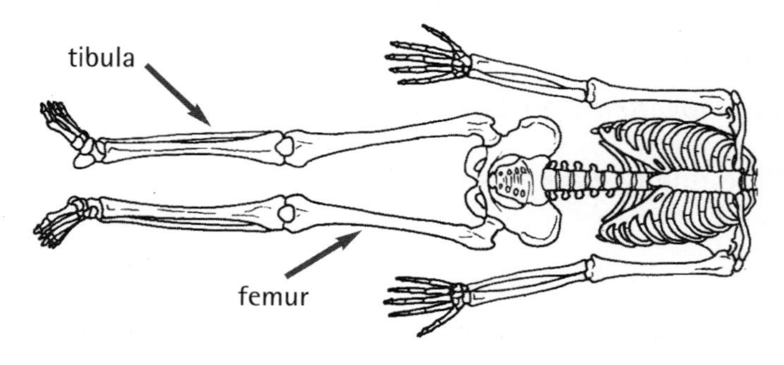

where h is the height of the person, f is the length of the femur and t is the length of the tibia

All the measurements are in centimetres.

REMEMBER The numbers in brackets are always done first. So put the numbers to be multiplied into brackets to remind you to do this next.

Find the height of a white male using the formula: $h = 1.3\ (f + t) + 63.29$, where $f = 44.3$ and $t = 34.9$. All measurements are in centimetres.

$$\begin{aligned} h &= 1.3\ (f + t) + 63.29 \\ &= 1.3\ (44.3 + 34.9) + 63.29 \\ &= (1.3 \times 79.2) + 63.29 \\ &= 102.96 + 63.29 \\ &= 166.25 \end{aligned}$$

So this person was approximately 166.25 cm tall when he was alive.

REMEMBER Always start by writing down the formula you are using.

REMEMBER Align your equal signs. It makes your answers neater.

Example 3

The following formula converts degrees Fahrenheit into degrees Centigrade.

$C = \frac{5}{9}\ (F - 32)$

where C is the temperature in degrees Centigrade and F is the temperature in degrees Fahrenheit.

Convert 59° Fahrenheit into degrees Centigrade.

$$\begin{aligned} C &= \tfrac{5}{9}\ (F - 32) \\ &= \tfrac{5}{9}\ (59 - 32) \\ &= \tfrac{5}{9} \times 27 \\ &= 27 \div 9 \times 5 \\ &= 15 \end{aligned}$$

So 59° Fahrenheit is equal to 15° Centigrade.

A rule that is found when working with a pattern is also called a formula. In the **Foundation Level** exam, you will probably be asked to find a rule. In the **General Level** exam, you are more likely to be asked to find a formula, but the meaning is exactly the same and you do exactly the same thing.

At **Foundation Level**, you can explain your rule or formula in words. At **General Level**, you must use algebra to describe the formula.

Practice question

1) The formula for the gradient of a straight line is $\frac{b - d}{a - c}$

   Find the gradient of a line when $a = 12$, $b = 7$, $c = 4$ and $d = 3$.

Multiplying out brackets and simplifying

When working with numbers, brackets are used to show which part of the sum to work out first.

Example 1

◎ *Multiply out the brackets 3(2 + 5)*

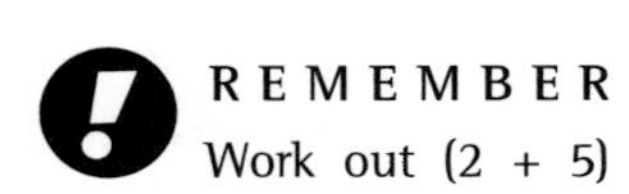

REMEMBER Work out (2 + 5) first.

$3(2 + 5)$

$= 3 \times 7$

$= 21$

Notice what happens to the 3 outside the brackets. The 3 multiplies the numbers inside the brackets.

This example can be worked out another way:

$3(2 + 5) = 3 \times 2 + 3 \times 5$

$= 6 + 15$

$= 21$

This time the 3 multiplies each of the numbers inside the brackets in turn. Notice that the answer is still the same.

When you use algebra, you should use this second method of working out.

Example 2

◎ *Multiply out the brackets 2(a + 7)*

$2(a + 7) = 2 \times a + 2 \times 7$

$= 2a + 14$

Example 3

◎ *Multiply out the brackets 5(y - 1)*

$5(y - 1) = 5 \times y - 5 \times 1$

$= 5y - 5$

If there are more terms, just remember that every term must be multiplied by whatever is immediately in front of the brackets.

REMEMBER Variables are usually written in alphabetical order, so *bm* is exactly the same as *mb*.

Example 4

◎ *Multiply out the brackets m(a - b + 3)*

$m(a - b + 3) = a \times m - b \times m + 3 \times m$

$= am - bm + 3m$

Simplifying expressions

When you have multiplied out brackets, you will sometimes have to do some tidying up. This tidying up is called 'simplifying'.

REMEMBER Only multiply the numbers that are inside the brackets.

Example 1

◎ *Multiply out the brackets and simplify $11(a + 5) - 8a$*

$$11(a + 5) - 8a$$
$$= 11 \times a + 11 \times 5 - 8a$$
$$= 11a + 55 - 8a$$
$$= 11a - 8a + 55$$
$$= 3a + 55$$

REMEMBER When you move terms around, the sign (+ or -) in front of the term moves too.

Example 2

◎ *Multiply out the brackets and simplify $2x(9 - y) + 5x$*

$$2x(9 - y) + 5x$$
$$= 2x \times 9 - 2x \times y + 5x$$
$$= 18x - 2xy + 5x$$
$$= 23x - 2xy$$

REMEMBER Tidy up $18x + 5x$

Factorising

Factorising is the process of putting algebra back into brackets by finding a common factor in every one of the terms. A factor is a number that divides into another number. When you are factorising, look for terms that divide exactly into the term you want to factorise.

Example 1

◎ *Factorise $3b - ab$*

? *What have $3b$ and ab got in common?*

Both terms have b in common, so: $3b - ab = 3 \times b - a \times b$
$= b(3 - a)$

Example 2

◎ *Factorise fully $3pq - 6q^2$*

REMEMBER 'Factorise fully' means that you have to find all the factors, not just one.

In this example, both the number and the variables (letters) have common factors. The numbers 3 and 6 have a common factor of 3. The variables pq and q^2 have a common factor of q. Putting the two factors together means that $3pq - 6q^2$ has a common factor of $3 \times q$ or $3q$ for short.

So: $3pq - 6q^2 = 3 \times p \times q - 2 \times 3 \times q \times q$
$= 3q \times p - 3q \times 2q$
$= 3q(p - 2q)$

REMEMBER q^2 means $q \times q$

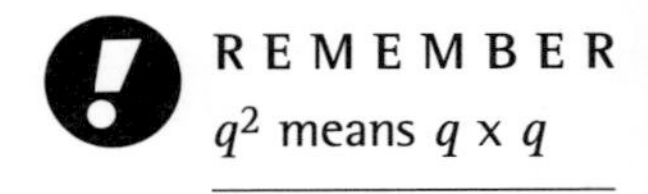

Solving equations

REMEMBER If the scales are to stay in balance, whatever you do to one side of the scales must be done to the other side as well.

Equations are just like a set of old-fashioned scales which have the same weight on both sides. Both sides of the equation are equal.

Example 1

In the picture on the right $3a + 1 = 22$

How do you find out what a is equal to?

Step 1: Write down the equation.

$$3a + 1 = 22$$

Step 2: Take 1 away from both sides.

$$3a + 1 = 22$$

$$\mathbf{-1} \qquad \mathbf{-1}$$

REMEMBER $3a$ means $3 \times a$

This gives: $3a = 21$

Step 3: Divide both sides by 3.

$$3a = 21$$

$$\div 3 \qquad \div 3$$

$$a = 7$$

Step 4: Check to make sure the answer is correct.

Check: $3a + 1 = (3 \times 7) + 1 = 21 + 1 = 22$

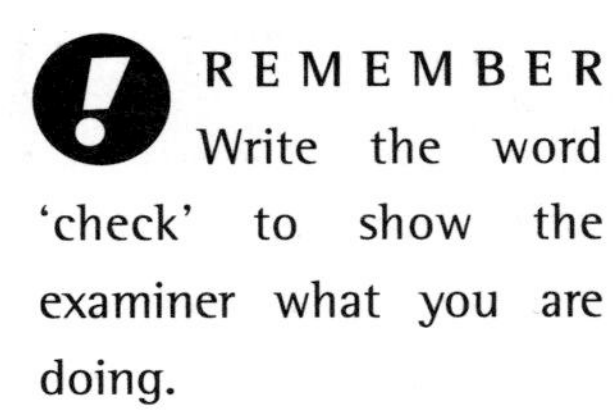

REMEMBER Write the word 'check' to show the examiner what you are doing.

The answer to the check is 22. This is the same as the right-hand side of the equation, so we know that $a = 7$ is definitely the right answer. Use this way of checking your answer in the exam to make sure you are right.

Example 2

Solve the equation $4y - 6 = 32$

REMEMBER You might have to find the value of x or another letter instead of y. It makes no difference to the way you solve the equation.

Step 1: $4y - 6 = 32$

Step 2: $+6 \qquad +6$ This time you have to add 6 to 'undo' the -6

$4y = 38$

Step 3: $\div 4 \qquad \div 4$ Divide both sides by 4 to find y

$y = 9\frac{1}{2}$

Step 4:

Check: $4y - 6 = (4 \times 9\frac{1}{2}) - 6 = 38 - 6 = 32$

This is the right-hand side, so $y = 9\frac{1}{2}$ is the correct answer.

Example 3

This is slightly more difficult, but you can use the same method. Make sure that you study the example carefully, until you understand what to do.

You need to get the variables (letters) on one side and the numbers on the other.

◎ *Solve the equation $6x - 3 = 2x + 5$*

Step 1:	$6x - 3$	=	$2x + 5$	
Step 2:	**+3**		**+3**	Add 3 to both sides to 'undo' -3
	$6x$	=	$2x + 8$	
Step 3:	**$-2x$**		**$-2x$**	Take $2x$ from both sides to 'undo' the $2x$
	$4x$	=	8	
Step 4:	**÷4**		**÷4**	Divide both sides by 4 to find x
	x	=	2	

Step 5: Check.

$6x - 3 = (6 \times 2) - 3 = 12 - 3 = 9$

$2x + 5 = (2 \times 2) + 5 = 4 + 5 = 9$

Both sides are equal when $x = 2$, so the answer is correct.

REMEMBER Show your working. It is important to let the examiner see how you have worked out the answers. **No working = no marks!**

Example 4

In this example, the steps have been shown without the written explanation. Make sure you understand them.

◎ *Solve the equation $4v + 1 = v + 40$*

Step 1:	$4v + 1$	=	$v + 40$
Step 2:	**-1**		**-1**
	$4v$	=	$v + 39$
Step 3:	**$-v$**		**$-v$**
	$3v$	=	39
Step 4:	**÷3**		**÷3**
	v	=	13

Step 5: Check.

$4v + 1 = (4 \times 13) + 1 = 52 + 1 = 53$

$v + 40 = 13 + 40 = 53$

So $v = 13$ is correct.

Solving inequalities

When things are equal in value, an equals sign is used. However, many things aren't equal. For example, Nisha is taller than Paul.

In maths we use four signs to represent the different kinds of inequality.

The first two signs are:

$>$ which means 'is greater than'

$<$ which means 'is less than'

The pointed end of the inequality sign is always directed at the smaller amount. So we could say:

Paul's height $<$ Nisha's height (Paul's height is less than Nisha's height)

or

Nisha's height $>$ Paul's height (Nisha's height is greater than Paul's height)

You can see that in both of these examples the pointed end of the inequality sign is directed at the smaller amount, in this case, Paul's height. In both examples, the wider end shows the larger amount.

Inequality signs can be used with numbers:

$3 < 5$	(3 is less than 5)	$5 > 3$	(5 is greater than 3)
$12 < 19$	(12 is less than 19)	$19 > 12$	(19 is greater than 12)

The other two signs are:

$\geq$ which means 'is greater than or equal to'

$\leq$ which means 'is less than or equal to'

The top parts of the signs are the same as the 'greater than' and 'less than' signs we just looked at. The bottom half of these signs are half of the equals sign. This is because they mean 'greater than **or equal to**' and 'less than **or equal to**'.

Look at this statement: $y \geq 6$

❓ *Which numbers could y stand for?*

y could be equal to 6 or 7 or 8 or more. In fact, y could be any number from 6 upwards.

When you solve an inequality, you do exactly the same as you do to solve an equation. The only difference is that you use the inequality sign instead of the equals sign.

Example 1

Solve algebraically $9d - 5 > 13$

Step 1:	$9d$	-5	> 13
Step 2:		$+5$	$+5$
	$9d$		> 18
Step 3:	$\div 9$		$\div 9$
Step 4:	d		> 2

Step 5: Check.

$9d - 5 = (9 \times 2) - 5 = 13 =$ the value on the right-hand side

So $d > 2$ is correct.

Even though you are working with inequalities, the check is the same as before. As long as the left-hand side of the inequality has the same value as the right-hand side, you have the correct answer.

Example 2

Solve algebraically $4h + 1 \leq h + 2$

Step 1:	$4h + 1$	$\leq$	$h + 2$
Step 2:	-1		-1
	$4h$	$\leq$	$h + 1$
Step 3:	$-h$		$-h$
	$3h$	$\leq$	1
Step 4:	$\div 3$		$\div 3$
Step 5:	h	$\leq$	$\frac{1}{3}$

Step 6: Check.

$4h + 1 = (4 \times \frac{1}{3}) + 1 = \frac{4}{3} + 1 = 1\frac{1}{3} + 1 = 2\frac{1}{3}$

$h + 2 = \frac{1}{3} + 2 = 2\frac{1}{3}$

so $h \leq \frac{1}{3}$ is correct.

WARNING

Credit Level students ONLY

There is only one exception to the general rule that you treat an inequality in exactly the same way that you would an equation.

If you have to multiply or divide both sides by a negative number, then the inequality sign changes to its opposite, for example:

$-8x \geq 56$

Divide both sides by -8 to give:

$x \leq -7$

REMEMBER Look back at pages 64–65 to remind you of the steps.

Using graphs

This section is about:

- working out the gradient of a straight line
- drawing and interpreting pie charts
- drawing and interpreting bar graphs and pictographs
- drawing straight line and curved line graphs
- working with time and distance graphs
- interpreting graphs

A graph is a sort of picture which is used to show information without having to use words to describe what is happening.

There are many types of graph and this section will look at the ones you need to know about for your **General Level** exam. At the beginning of the **General Level** exam you are given a formulae list. In this list you will find the following formula for finding the gradient of a straight line.

$$\text{Gradient} = \frac{\text{vertical height}}{\text{horizontal distance}}$$

Look at the FactZONE opposite and at pages 70 and 71 for more information on this formula.

You should have the necessary equipment with you in the exam in case you have to draw any graphs. You will need a sharp pencil (or two), a pencil sharpener, a rubber, a ruler, compasses (for drawing circles) and a protractor or angle measurer. Most exam centres will have the last two or three items available for you to borrow, but it is better to use equipment you are used to.

If you are asked to **draw** a graph, you will have to be as accurate as possible, probably working out some values and plotting them, then joining up the line.

If you are asked to **sketch** a graph, then you are being asked for an interpretation or estimate of what you think the graph should look like.

For any graph, think about what information you have or you need to find before you start. Then be as neat and accurate as you can.

FactZONE

Gradient

The formula for working out the gradient of a line on a graph is:

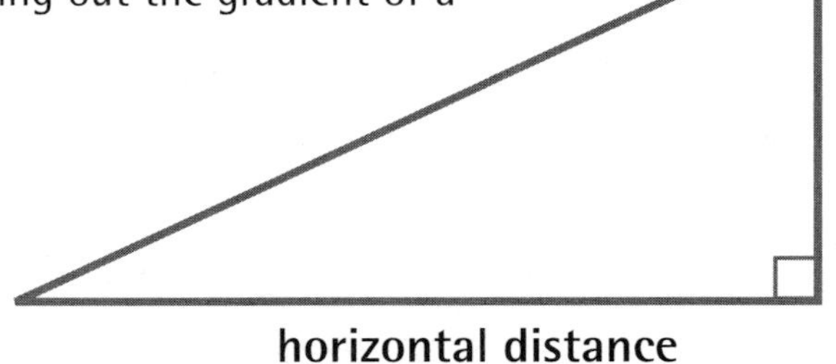

$$\text{Gradient} = \frac{\text{vertical height}}{\text{horizontal distance}}$$

Some useful facts about graphs

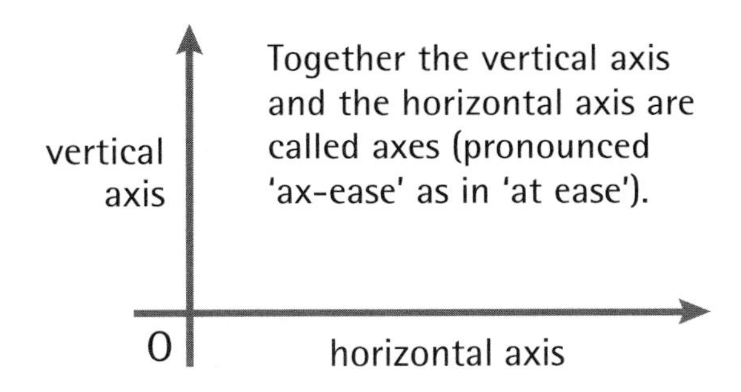

Where the axes cross is called the 'origin', often marked with a zero or an 'O' to stand for origin. The origin has coordinates (0, 0).

The scale is a set of numbers you write along the axes. They do not have to be the same on both axes, but on each axis they must go up in equal steps, for example:

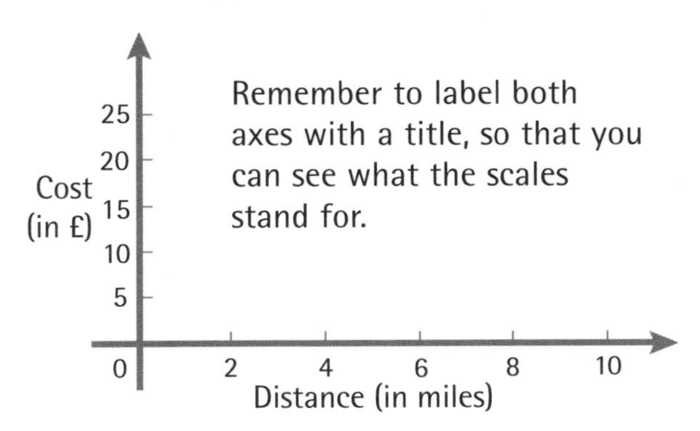

Gradient

Gradient is a measure of steepness.

You may have seen a gradient sign as you were travelling up or down a steep hill. Gradient signs warn you how steep the slope is so that you can slow down.

On the exam paper, gradient is given on the formulae list as:

$$\textbf{Gradient} = \frac{\textbf{vertical height}}{\textbf{horizontal distance}}$$

If you want to work out the gradient of a line, you divide the vertical height by the horizontal distance.

Example 1

Work out the gradient of this slope.

23 cm

92 cm

> **REMEMBER**
> The units of measurement that you divide must be the same, centimetres divided by centimetres or metres divided by metres, and so on.

gradient = vertical height ÷ horizontal distance

= 23 ÷ 92

= 0.25

The gradient of the slope is 0.25.

There are no units of measurement for the gradient. It is just a number representing how steep the slope is.

A horizontal gradient has a gradient of zero (it has no slope). The higher the number, the steeper the slope.

Example 2

Plot the points (-7, -4) and (8, 5). Join them to make a straight line. What is the gradient of the line?

Plot the points on squared paper. Join them up (with a ruler) to make a straight line.

The sloping line has to become the hypotenuse of a right-angled triangle. Draw in two more straight lines along the grid to make this a right-angled triangle. Look at the illustration at the top of the page opposite.

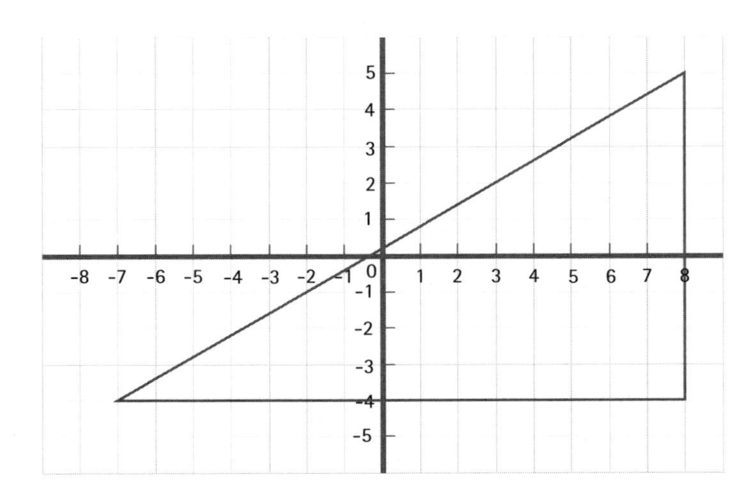

REMEMBER Look back to page 40 if you need more information on plotting coordinates and to page 42 if you need to find out about right-angled triangles.

Count the squares: vertical height = 9 squares

horizontal distance = 15 squares

gradient = vertical height ÷ horizontal distance

= 9 ÷ 15

= 0.6

REMEMBER You can show a gradient as a fraction or as a decimal. If you do use a fraction, make sure it has been simplified. For more on using fractions, see page 16.

Example 3

In this example, the vertical height is negative. This is because the 'height' is actually a 'drop'.

The horizontal distance is still a positive number.

So the gradient = -6 ÷ 9

= $-\frac{2}{3}$ or -0.67

A line that slopes down from left to right has a negative gradient.

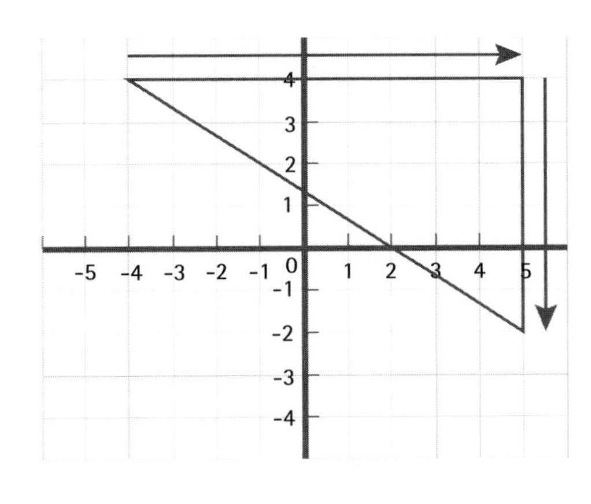

REMEMBER If the line goes up from left to right, the gradient is positive. If it goes down, it's negative.

Practice question

1) Draw a horizontal axis from -10 to 10 and vertical axis from -10 to 10.

Plot the following points and find their gradient.

a) (0, -5) and (5, 0)

b) (-8, 4) and (7, 4)

c) (-7, 9) and (1, -3)

Pictographs and bar graphs

Pictographs

A pictograph is exactly what the name tells you – it's a graph that uses pictures to represent information.

Example

A national building firm produces a pictograph for an advertising leaflet to show the number of houses it has built over the last five years.

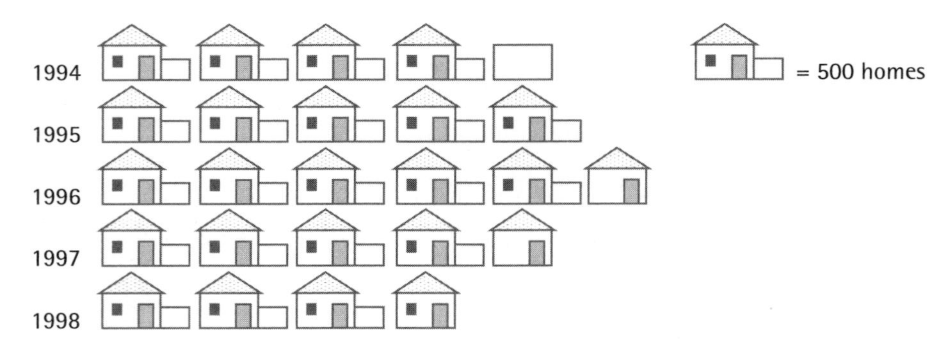

In 1996, the firm built 2800 houses.

Explain why *stands for 300 houses.*

Look at the 1996 line. You know that the picture stands for 500 houses and you are told that in 1996 the firm built 2800 houses. In the 1996 line there are five complete pictures and one part picture. The five complete pictures stand for 5 x 500 or 2500 houses, so the part picture must be 300 houses to make a total of 2800 houses.

How many houses has the firm built altogether over the last five years?

First you have to work out what the other pictures stand for.

stands for 500 houses. It has five parts: building, roof, door, window and garage. So each separate part must be equal to 100 houses.

Check this with the picture for 300 houses: building, roof and door make three parts, so 300 houses.

Now you can work out each line:

The firm has built 2100 + 2500 + 2800 + 2300 + 1900 = 11 600 houses over last five years.

Describe fully what the pictograph shows.

The pictograph shows that starting with 2100 houses in 1994, the number of houses built rose to a maximum in 1996, when the firm built 2800 houses. Since then building has declined to 1900 houses in 1998.

REMEMBER When you are asked to describe fully what a pictograph shows, make sure that you use words such as 'maximum' and 'minimum' and give relevant numbers to back up your answer.

Bar graphs

Bar graphs are what they say they are – graphs with bars on!

Example

The table below shows the distance of five planets in our solar system from the Sun.

Planet	Distance from the Sun (to the nearest million km)
Mercury	58 000 000
Venus	108 000 000
Earth	150 000 000
Mars	228 000 000
Jupiter	778 000 000

On a sheet of squared paper, draw a bar graph to show this information.

REMEMBER In the exam, there would be a squared grid on the paper for you to draw the bar graph.

Step 1: Decide the scale.

You need draw a vertical axis to start at zero and go up to the greatest distance, 778 million km. The easiest way to think of the scale is to think of all the numbers as so many million kilometres. Then you just have to plot 58, 108, 150, 228 and 778.

For this example, the easiest scale is to go from zero to 800 in steps of 100, moving up one square at a time. This means that your graph is 8 squares high. This is a good way of coping with the very large numbers.

Label the vertical axis 'Distance in millions of km'.

Step 2: Draw the bars.

You will have to estimate where to stop drawing the bar as only Earth, at 150 million km, will stop exactly on a line. The bars should all be the same width.

REMEMBER Use a ruler, count carefully and be as accurate as possible.

Label the horizontal axis with the planet names and a title. Your bar graph should look like this:

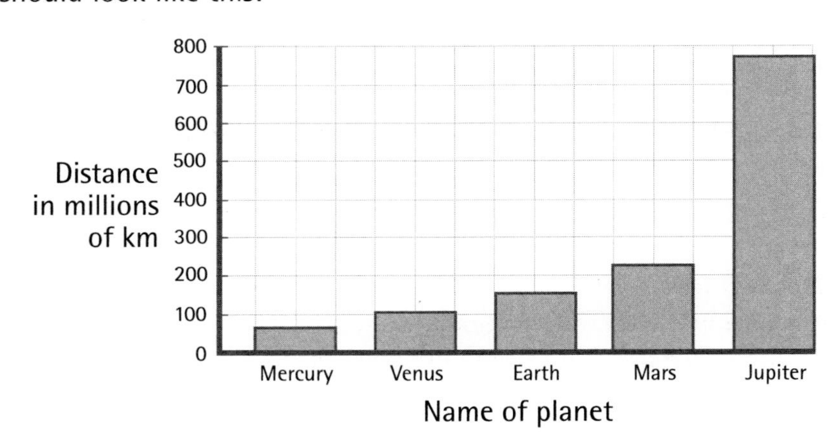

Pie charts

Pie charts are circular diagrams – the shape of a pie – split up into 'slices'. A pie chart shows numerical information in picture form, the 'slices' show how this information is divided up.

REMEMBER The angles around a point = 360°.

The angle at the centre of the circle is 360°.

If you know the total amount the whole pie chart represents and the angle size of the slices, you can work out how much each slice represents.

Example 1

This pie chart shows the running costs of a car last year. The total amount spent was £2160.

repairs
road tax
140°
petrol

◎ *Calculate the amount spent on repairs last year.*

REMEMBER This is because there are 360° in a full turn and so there are 360° at the centre of the circle.

Step 1: Write down the total amount spent.

£2160

Step 2: Multiply this by the angle at the centre and divide by 360.

2160 x 140 ÷ 360 = £840

So £840 was spent on repairing the car last year.

◎ *Calculate how much was spent on road tax last year.*

Use your protractor to measure the size of the angle representing road tax. It is marked with a small arc like this:

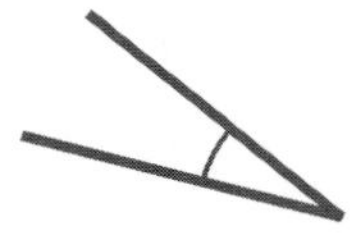

Did you measure the angle to be about 25°?

Following the steps above, use 25° to work out the road tax:

2160 x 25 ÷ 360 = £150

◎ *How much did the petrol cost?*

Subtract the two amounts you know from the total of £2160 to get the third amount, the amount spent on petrol.

2160 - 840 - 150 = £1170

Example 2

Ian has just come back from a week's holiday in Ireland.

His fare cost £90, the hotel cost £378 and he spent £180 while he was there.

◎ *Draw a pie chart showing how much he spent on his fare, the hotel and spending money.*

Step 1: Work out the total amount of money that Ian spent.

£90 (fare) + £378 (hotel) + £180 (spending money) = £648

Step 2: Work out the angle at the centre of the pie chart for each of the three categories.

This is like working out a percentage, but instead of multiplying by 100 (for 100% or the whole amount) in the calculation, multiply by 360 instead.

Divide the amount spent in each category by the total amount and then multiply by 360.

fare = $90 \div 648 \times 360 = 50°$

hotel = $378 \div 648 \times 360 = 210°$

spending money = $180 \div 648 \times 360 = 100°$

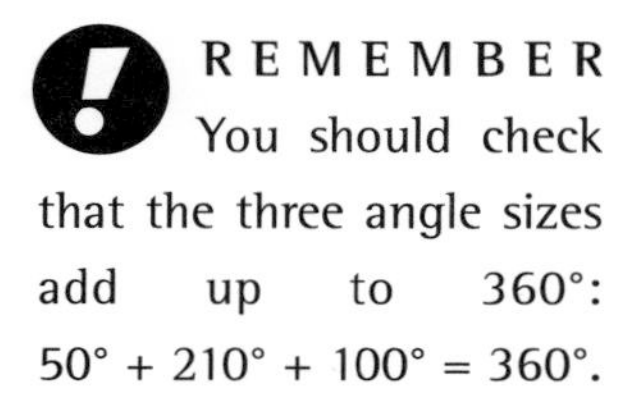
REMEMBER You should check that the three angle sizes add up to 360°: 50° + 210° + 100° = 360°.

Step 3: Draw the pie chart.

First draw a circle and then a radius.

Measure 50° from the radius and draw in another radius. Mark the slice 'fare'.

It is easier with a protractor to measure 100° than 210°, so do the 100° angle next. It can be measured from either of the existing radius lines. Label this slice 'spending money'.

Check that the slice you have left measures 210°. Label this slice 'hotel'.

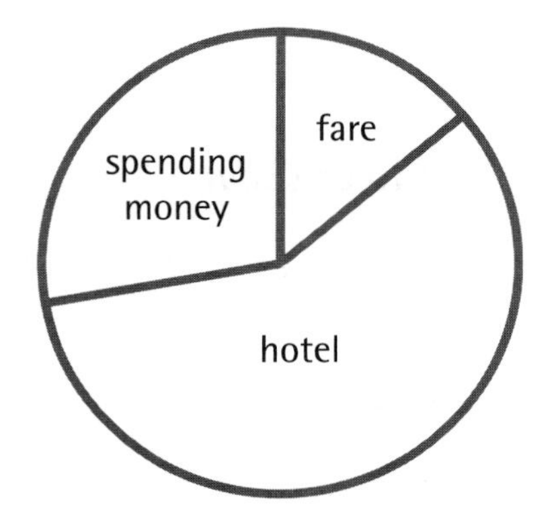

Drawing straight line graphs

Every graph of a straight line has an equation which looks something like these:

$y = 5x$ or $y = 3x - 2$ or $y = x + 6$

Example 1

Draw the graph of $y = 3x - 2$.

Step 1: Draw a table of values, so that you can work out the coordinates to plot on the graph.

You only need two points to draw a straight line, but it is sensible to use three points because you have more chance of correcting any mistakes you might make.

Choose easy values to use to stand for x. You might be given the values you have to use in the exam. If you aren't, $x = 1$, $x = 2$ and $x = 3$ are easy to work out. Make a table like this:

x	1	2	3
y			

Step 2: Work out the values of y to go into the table.

The equation is $y = 3x - 2$, so:

when $x = 1$, $y = (3 \times 1) - 2 = 3 - 2 = 1$ so 1 goes in the table underneath $x = 1$

when $x = 2$, $y = (3 \times 2) - 2 = 6 - 2 = 4$ so 4 goes in the table underneath $x = 2$

when $x = 3$, $y = (3 \times 3) - 2 = 9 - 2 = 7$ so 7 goes in the table underneath $x = 3$

x	1	2	3
y	1	4	7

REMEMBER Coordinates have an x and a y value like this: (x, y). Look back to page 40 for how to plot coordinates.

REMEMBER Draw the axes longer than you need and remember to write in the scales.

Step 3: If they aren't already drawn for you, draw the axes.

The horizontal axis is for the x values, the vertical axis is for the y values.

Step 4: Plot the three pairs of coordinates from the table, using dots (not crosses)!

The coordinates are (1, 1), (2, 4) and (3, 7). Join them up with a straight line. Draw the line long enough to fill the whole graph. Now label the graph with the 'name' of the equation.

This is what your graph should look like.

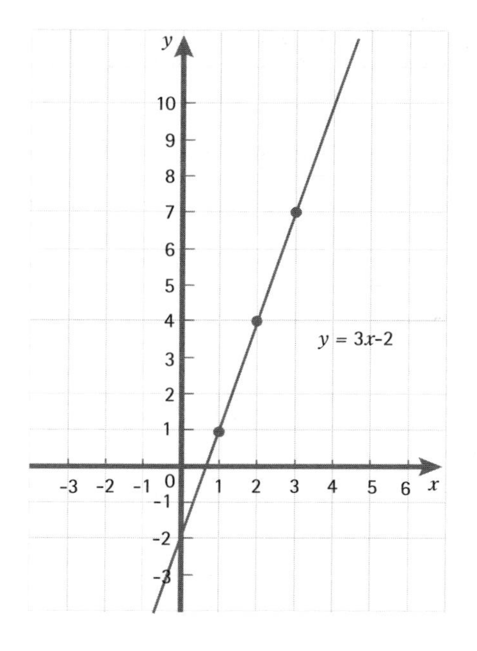

REMEMBER It's easier to draw graphs with a long ruler.

Example 2

It costs a record company £52 000 to make the master disc for a CD and then £6 per disc (or £6000 for each 1000 discs) after that. The record company sells the CDs for £10 each. In other words, it gets £10 000 for every 1000 CDs it sells.

This table of values shows costs and potential sales.

No of CDs (in thousands)	1	2	3	5	10	15
Cost in £1000s	58	64	70	82	112	142
Sales revenue in £1000s	10	20	30	50	100	150

What is the break-even point?

The break-even point is when a company has received enough money from sales to pay for all the costs, but hasn't made any profit yet. On the graph it is where the two lines cross.

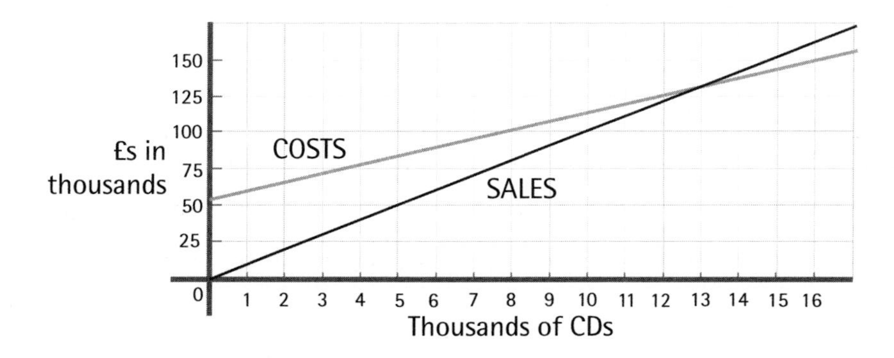

The lines cross at 13 on the 'Thousands of CDs' axis. So, the record company must sell 13 000 CDs to break even, and more than 13 000 CDs to start making a profit.

REMEMBER Don't try to use shortcuts when you're drawing a graph. Be as accurate as you can.

Drawing curved line graphs

If you have to draw a curved line graph, you will be given more information than for a straight line graph.

Example

A box is being designed with a square base. The box is to be 10 cm high.

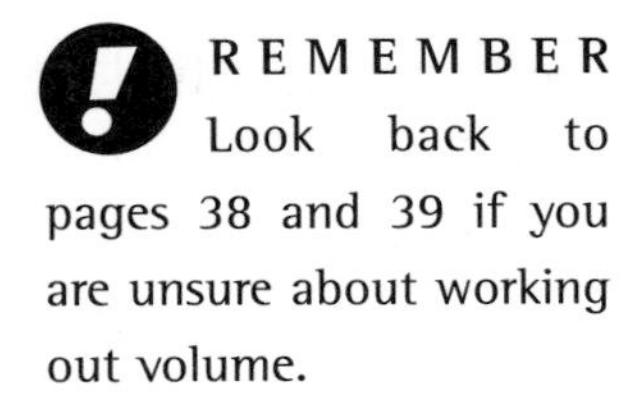

REMEMBER Look back to pages 38 and 39 if you are unsure about working out volume.

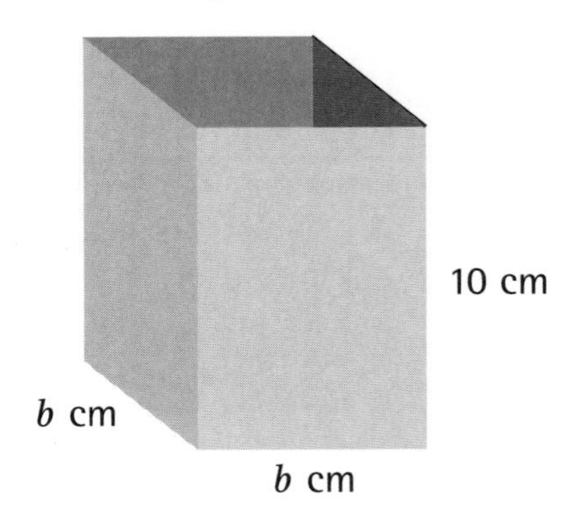

REMEMBER b^2 means $b \times b$ not $b \times 2$

The area of the base is b^2. So the volume of the box is $b^2 \times 10$ or $10b^2$.

Draw the curved graph for the box above.

Step 1: Work out the volume of the box for each of the gaps in the table below. Put the answers into the table.

Length of base in cm	0	1	2	3	4	5	6	7
Volume of box in cm³	0	10	40			250		

For a base length of 3, volume $= 10 \times b \times b = 10 \times 3 \times 3 = 90$

For a base length of 4, volume $= 10 \times b \times b = 10 \times 4 \times 4 = 160$

For a base length of 6, volume $= 10 \times b \times b = 10 \times 6 \times 6 = 360$

For a base length of 7, volume $= 10 \times b \times b = 10 \times 7 \times 7 = 490$

The table will now be:

Length of base in cm	0	1	2	3	4	5	6	7
Volume of box in cm³	0	10	40	*90*	*160*	250	*360*	*490*

Step 2: The axes will be drawn for you in the exam, but draw them on graph paper now so that you can practise.

Draw the horizontal axis from 0 to 8 units, going up in steps of 1. Label the axis 'Length of base in cm'.

Draw the vertical axis from 0 to 500, going up in steps of 50. Label the axis 'Volume of box in cm³'.

Step 3: Draw the graph. In the exam the beginning of the graph may be drawn in for you to continue.

Plot and join up the pairs of coordinates.

The coordinates are: (0, 0), (1, 10), (2, 40), (3, 90), (4, 160), (5, 250), (6, 360) and (7, 490).

Join them up with a smooth curved line (not a straight line).

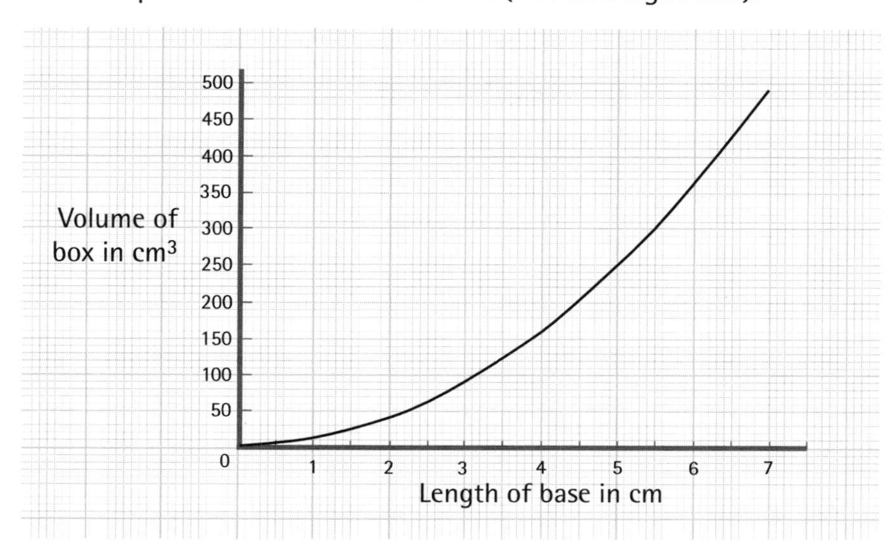

REMEMBER Practise drawing freehand curved lines which go through dots marked on graph paper.

What is the volume of the box when the length of the base is 6.5 cm?

Read from the graph. Find the point on the horizontal axis where the base is 6.5 cm (half way between 6 and 7). Follow that line up until it meets the graph, then read off the value on the vertical axis. The volume is about 420 cm^3.

What is the length of the base when the volume of the box is 200 cm^3?

This time start on the vertical axis at the point where the volume equals 200 cm^3. As before, follow the line along until it meets the graph and then read off the value on the horizontal axis. When the volume is 200 cm^3, the length of the base is about 4.5 cm.

Practice question

1) Draw a graph using the values in the following table.

What is the value of x when $y = 12$?

x	0	1	2	3	4	5
y	2	1	2	5	10	17

Time and distance graphs

Time and distance graphs often involve reading information from the graph as well as understanding what it means when two lines cross each other.

Example

Inverness and Dyce Airport (just outside Aberdeen) are 100 miles apart.

Matthew leaves Dyce in the Supportair van at 9 am. He travels at a steady speed of 50 miles per hour. 25 miles from Dyce he has to stop when the van has a flat tyre. After changing the tyre, he continues to Inverness at the same steady speed as before. The graph of his journey is shown below.

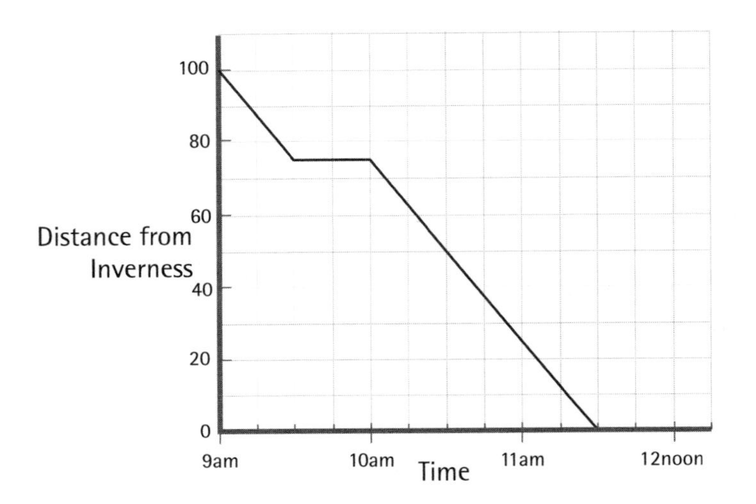

◎ *How long does Matthew stop to change the tyre?*

The horizontal line on the journey line shows the time that the van was stopped. This is half an hour.

◎ *When does Matthew arrive in Inverness?*

Reading from the graph, the line representing the journey stops half way between 11 am and 12 noon. This means he arrives at 11.30 am.

Kirsty sets off from Inverness in her car at 9 am. She travels at a steady speed of 60 miles per hour.

◎ *Draw Kirsty's journey on the same graph.*

Kirsty's journey line is a straight line from Inverness because she travelled the whole distance at the same speed. Notice the difference between the journey Matthew made and the journey Kirsty made. Matthew starts from Dyce and so his journey line must begin there. Kirsty begins her journey in Inverness.

REMEMBER Take care when you are working with time and distance graphs, allow yourself time to study the graph carefully, so that you understand what is happening.

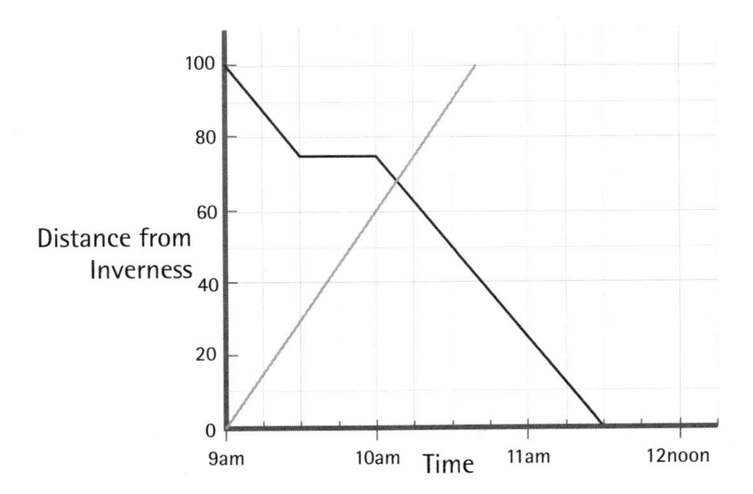

When do the two vehicles pass each other? How far are they from Inverness?

The two journey lines cross at about 10.10 am about 68 miles from Inverness.

Interpreting time and distance graphs

On a time and distance graph, the steeper the line, the faster the vehicle or person is moving.

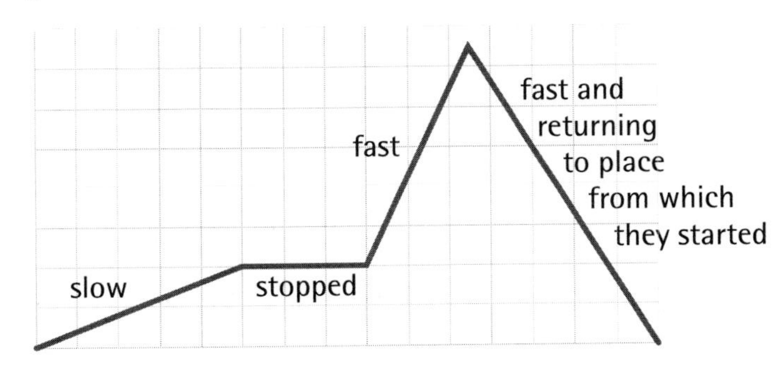

A line going down from left to right, as on the right-hand side of this diagram, means that the vehicle or person is returning to the place from which they started.

REMEMBER Use your common sense as well as your maths knowledge when interpreting graphs.

Interpreting graphs

Being able to draw a graph is just one step. The next process is to use it to answer questions.

Example 1

The graph below shows a rural school bus travelling from school to its last drop-off point at the top of a glen.

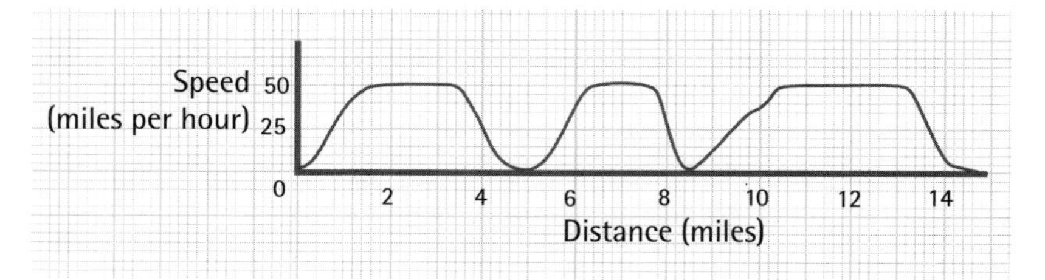

> **REMEMBER** If you are asked a question such as 'Does the bus go faster than 50 mph?', say 'Yes' or 'No' and give a reason. A good rule is to give one more fact than there are marks for the question, so if the question is for 2 marks, give 3 facts if you can. Be specific!

◎ *What is the top speed of the bus?*

The graph reaches the 50 line on the vertical or speed axis. So the top speed that the bus reaches is 50 miles per hour.

◎ *After how many miles does the bus make its first stop?*

It stops when the speed is zero mph. This happens first 5 miles from the school.

◎ *How many stops does the bus make?*

The bus stops three times. The beginning of the graph can't be counted as a stop, as this is when the pupils get on and the journey begins.

◎ *How far is it from school to the top of the glen?*

The total length that the graph covers on the horizontal axis is 15 miles. So it is 15 miles from school to the top of the glen.

Example 2

The graph below shows the average rainfall for the different seasons over four years.

Sp = Spring
Su = Summer
A = Autumn
W = Winter

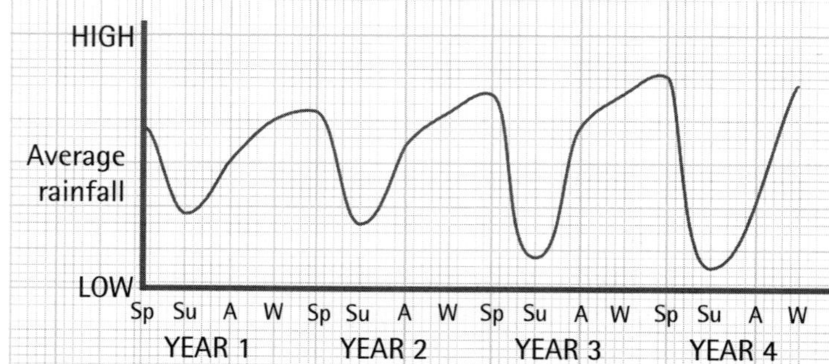

◎ *Does the graph show a changing weather pattern? Describe the general trend of the graph.*

? *What do you notice about the graph? What is happening to the amount of rain falling each year?*

The rainfall pattern is becoming more extreme, the peaks on the graph are getting higher and the dips are getting lower.

Look at the seasons on the horizontal axis. Place the ruler horizontally across the graph to examine each season in turn. Spring each year is when the maximum values appear. As the maximum is a little higher each year, this shows that springs are becoming wetter. Summers have the lowest rainfall each year and are becoming drier. As there are no numbers on the rainfall axis, it is not possible to use numbers in the answer. So a good answer might be something like this:

'Yes, the graph does show a changing weather pattern. Each year autumn, winter and spring have a little more rainfall than the year before, but summer is getting a little less. So summers are becoming drier and the other three seasons are getting wetter.'

REMEMBER 'Describe the general trend' means describe in words what the graph is telling you.

REMEMBER You need to relate the information to what you know about the real world.

Practice question

1) Donna sets out for school by bus at 8 am. The bus journey is shown on the graph on the right.

   Her friend Sandra lives next door. She cycles to school and leaves at 8.10 am. It takes her 35 minutes to get to school. Copy the graph and draw Sandra's journey onto it.

   At what time do the bus and bicycle pass each other?

   Who gets to school first, Donna or Sandra and by how many minutes? Why do you think that this might have happened?

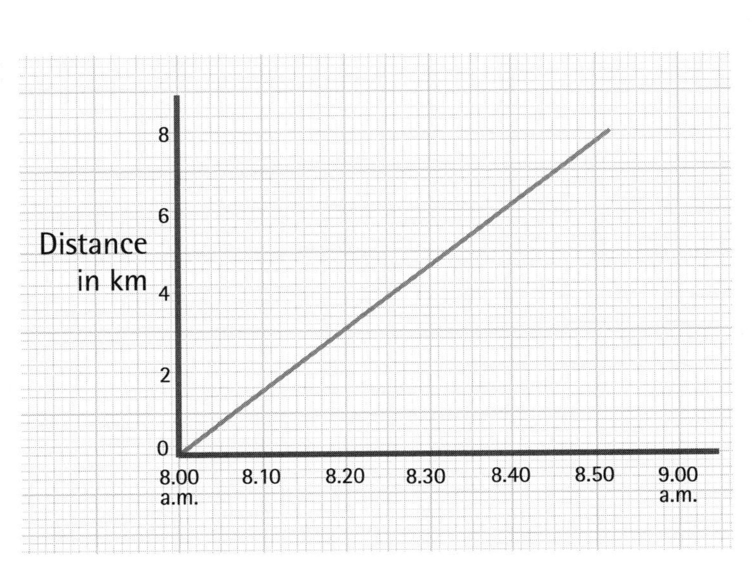

Answers: practice questions

Section 1: Statistics

Statistics (page 10)

1) mean = (25 + 26 + 39 + 49 + 52 + 57 + 59 + 61 + 61 + 62) ÷ 10 = 491 ÷ 10 = 49.1 cm; median = (52 + 57) ÷ 2 = 109 ÷ 2 = 54.5 cm; mode = 61 cm; range = 62 - 25 = 37 cm

Probability (page 11)

1) There are 52 cards in a normal pack of playing cards, 4 of them are Kings. So P(King) = $\frac{4}{52}$ = $\frac{1}{13}$ = 0.8

2) There are four different suits in a normal pack of playing cards – hearts and diamonds (which are red) and clubs and spades (which are black). There are 13 cards in each suit.
So P(heart) = $\frac{13}{52}$ = $\frac{1}{4}$ = 0.25

Section 2: Using number

Fractions and percentages (page 18)

1) 64 ÷ 8 x 5 = £40
2) 19 000 x 3.5 ÷ 100 = £665; 665 ÷ 12 x 8 = £443.33
3) 115.8 x 3 ÷ 100 = 3.474 = £3.47; 115.80 + 3.47 = £119.27

Averages (page 19)

1) 794 ÷ 12 = 66.17 = 66
2) 63.4 x 14 = 887.6; 64.12 x 15 = 961.8; 961.8 - 887.6 = 74.2 kg

Foreign exchange (page 20)

1) 450 x 1.61 = $724.50
2) 80 ÷ 3.2 = £25.00; £25 - £2 = £23.00

Time, distance and speed (page 24)

1) distance = speed x time = 65 x 4.5 = 292.5 km
2) time = distance ÷ speed = 560 ÷ 452 = 1.238; minutes = 0.238 x 60 = 14.43 = 14 minutes (to the nearest minute), so time taken is 1 hour 14 minutes

Reading two-way tables (page 25)

1) 2 hours 24 minutes
2) distance = 105 miles; minutes = $24 \div 60 = 0.4$, so total time is 2.4 hours; speed = distance ÷ time = $105 \div 2.4 = 43.75$ miles per hour

Section 3: Shape and space

Compass points and angles (page 29)

1) a) $90° - 30° = 60°$
   b) your scale drawing
   c) $38° \pm 2°$ – this means between 36° (38° - 2°) and 40° (38° + 2°)
   d) bearing = 180 + 90 + 38 = 308°

The triangle (page 33)

1) Area $= \frac{1}{2}$ base x height
   $= \frac{1}{2} \times 4.2 \times 2.5$
   $= 2.1 \times 2.5 = 5.25 \text{ cm}^2$

2)

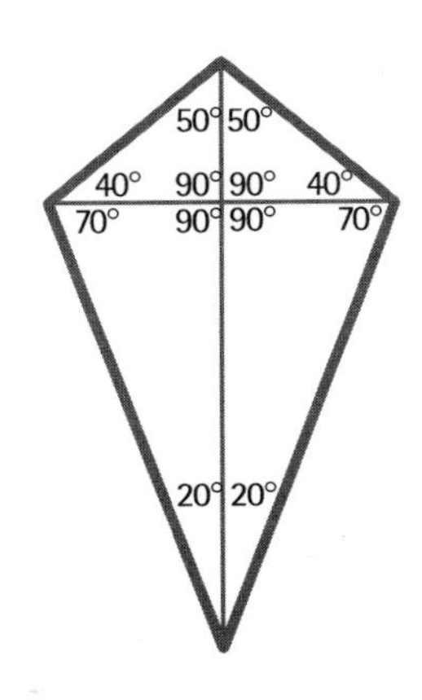

The circle (page 35)

1) Area of whole circle = $\pi r^2 = \pi \times 45^2 = 6361.7 \text{ cm}^2$
   $\frac{3}{4}$ of circle = $6361.7 \div 4 \times 3 = 4771.3 \text{ cm}^2$
   Area of rectangle = $85 \times 45 = 3825 \text{ cm}^2$
   so total area = $4771.3 + 3825 = 8596.3 \text{ cm}^2$

Section 5: Algebra

Indices and scientific notation (page 57)

1) $5^6 = 5 \times 5 \times 5 \times 5 \times 5 \times 5 = 15\,625$
2) $0.000\,054\,5 = 5.45 \times 10^{-5}$
3) 7.5 million = $7\,500\,000 = 7.5 \times 10^6$
4) $8.4 \times 10^{-3} = 0.0084$

Using formulae (page 61)

1) The formula is $\frac{b - d}{a - c}$ so $\frac{7 - 3}{12 - 4} = \frac{4}{8} = \frac{1}{2}$

Section 6: Using graphs

Gradient (page 71)

1) a) $5 \div 5 = 1$

b) $0 \div 15 = 0$

c) $-12 \div 8 = -1\frac{1}{2}$ or -1.5

Drawing curved line graphs (page 79)

1)

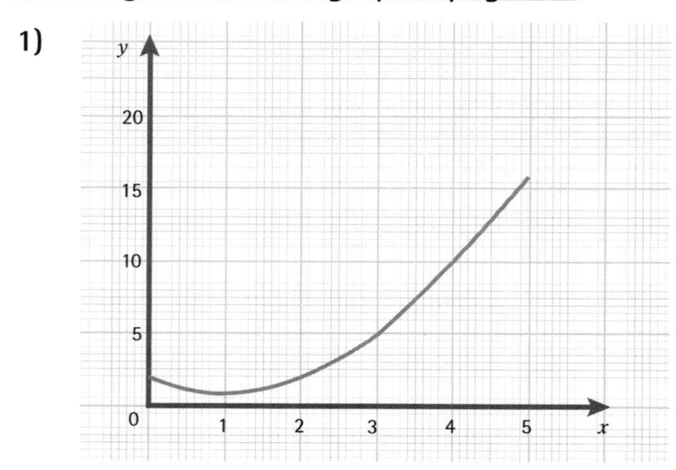

$x = 4.3$ (approximately)

Interpreting graphs (page 83)

1) The bus and bicycle pass each other at 8.31 am.

Sandra on her bicycle gets to school first by about 7 minutes.

This might be because the bus got stuck in traffic while Sandra was able to keep going, perhaps using cycle lanes. (Any sensible answer would be acceptable for this type of question, but make sure you give realistic reasons.)

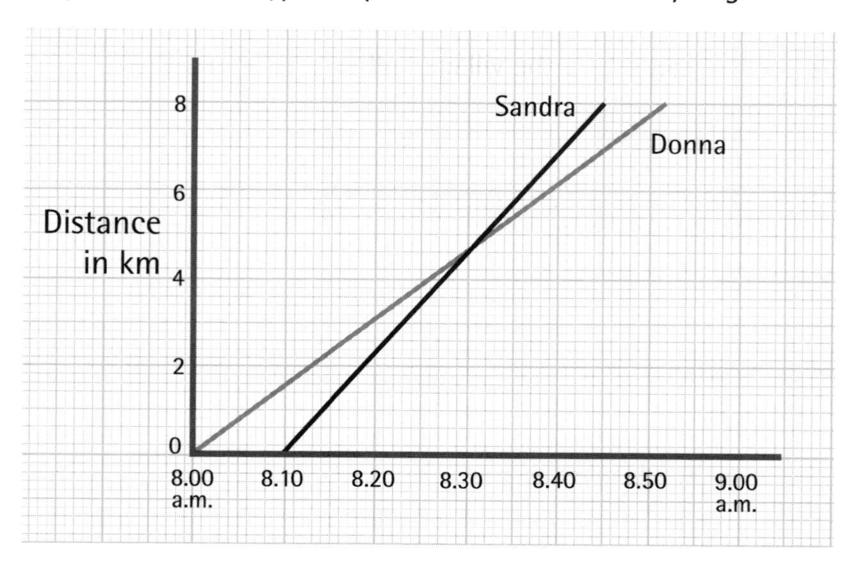

Exam-style questions

Section 1: Statistics

Probability

1) Jan and Rob are playing a game with an eight-sided spinner, with sides numbered 1 to 8. What is the probability of the spinner landing on a number greater than 5?

Section 2: Using number

Insurance and hire purchase

2) Look at the table on page 15.

   How much would it cost to insure a building for £65 000 in Area 3?

3) Using the same table, how much would it cost to insure the contents of a house for £2100 in Area 1?

4) James buys a three piece suite on interest-free credit (in other words, he pays no interest on the amount of money he has borrowed).

   The suite costs £995.00.

   He has to pay a 20% deposit and pay the balance in 10 equal payments.

   How much is each payment?

Ratio and proportion

5) Red, yellow and black dyes are mixed to dye fabric in the ratio of 6:2:1. If 48 grams of red dye are to be used, how much yellow and black dye is needed?

6) Normally 20 people can set up the village hall for the Shinty Club Sale of Work in 30 minutes. If only 15 people can help, how long will it take them?

7) During a flu epidemic $\frac{4}{7}$ of 427 pupils were off school.
   How many pupils were off school?

Time

8) A zoo opened on 21st April and closed again on the last day in September.

   It had a total of 249 064 visitors during that time.

   What was the average number of visitors per day?

Section 3: Shape and space

Area and perimeter

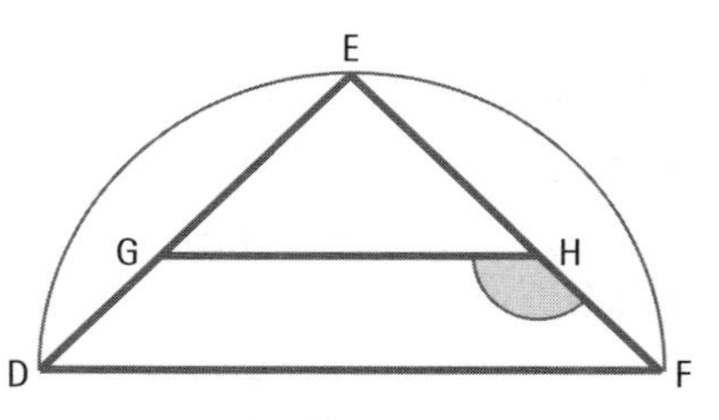

9) This drawing shows the arch of an old bridge which is in the shape of a semicircle. In order to strengthen it, engineers have put in two metal supports DE and EF, both the same length.

Later they have to add a horizontal support GH. GH is parallel to DF.

a) What is the size of angle DEF?

b) Calculate the size of the shaded angle GHF.

10) Find the area of this kite.

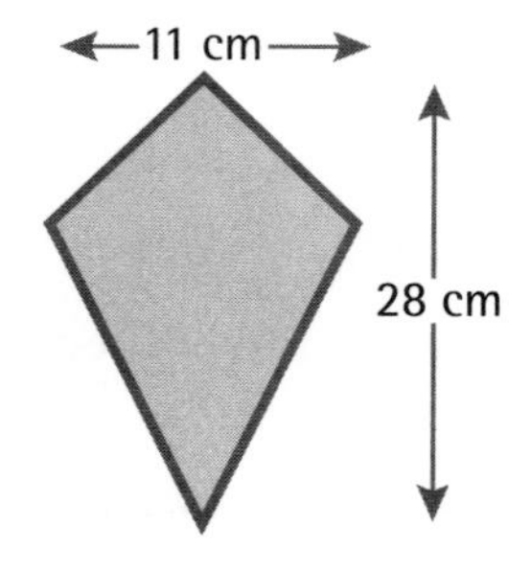

11) Find the distance round this running track. It has a semicircle with a diameter of 38 m at each end. The straights are 140 m long.

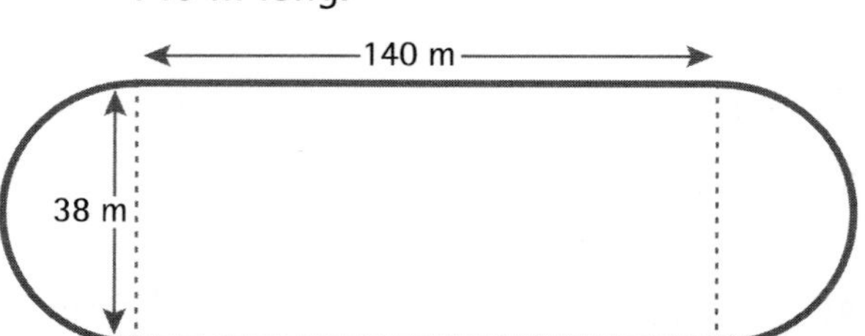

Volume

12) This can has a radius of 1.6 cm and a height of 5.5 cm. How many millilitres of tomato juice does it hold?

Coordinates and symmetry

13) Draw coordinate axes from -9 to 9 on the horizontal axis and from -6 to 6 on the vertical axis.

a) Plot the following points, joining each point to the previous point with a straight line:

(0, 0), (-2, 3), (-2, 5), (-5, 5), (-8, -4), (-2, -3), (0, 0)

b) Using the vertical axis as the line of symmetry, draw the reflection of the design from a).

c) Starting at (0, 0) and working clockwise, write down the coordinates of all points in the reflected drawing.

d) Copy the original image onto new axes. Now draw a rotation through 180° centred on (0, 0). What are the coordinates this time?

Section 4: Right angled triangles

Pythagoras' theorem

14) Kate is building a ramp for snake-boarding. What is the total length of plywood she needs to buy to cover both ramps and the whole top surface?

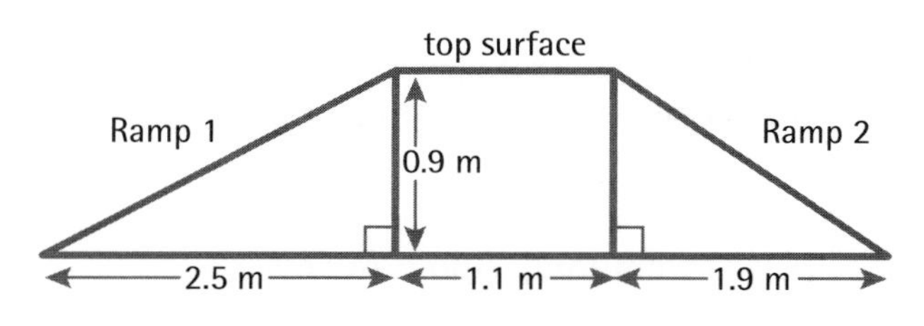

15) Jim wants to fit a wheelchair ramp to his house. It needs to be 60 cm high. The ramp is 250 cm long. Calculate the length of ground the ramp will need, marked *d* cm on the diagram.

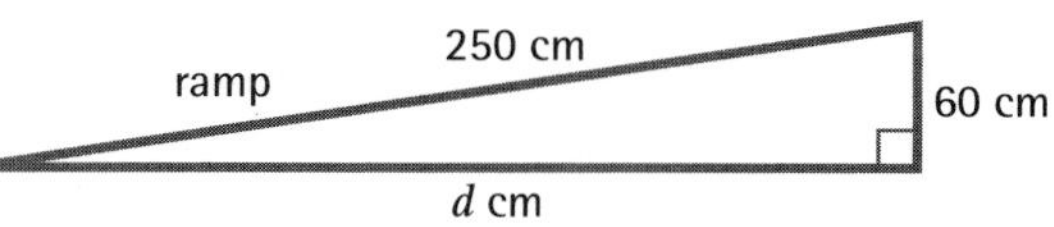

Trigonometry

16) Calculate the size of the angle, marked $x°$, at the base of the ramp. For safety, the angle should not exceed 15°. Is the ramp safe?

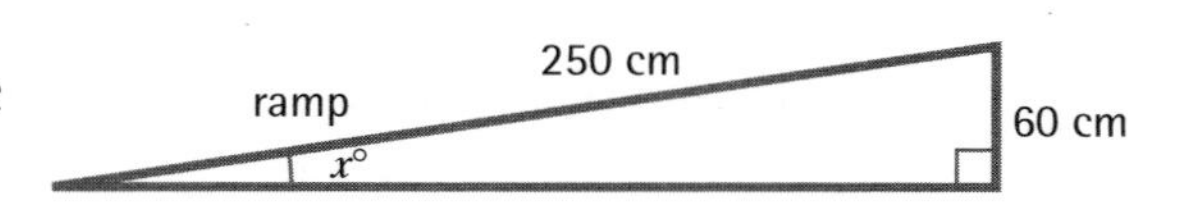

17) An architect has designed a roof. She knows the measurements shown.

a) Calculate the height, *h*, of the roof.

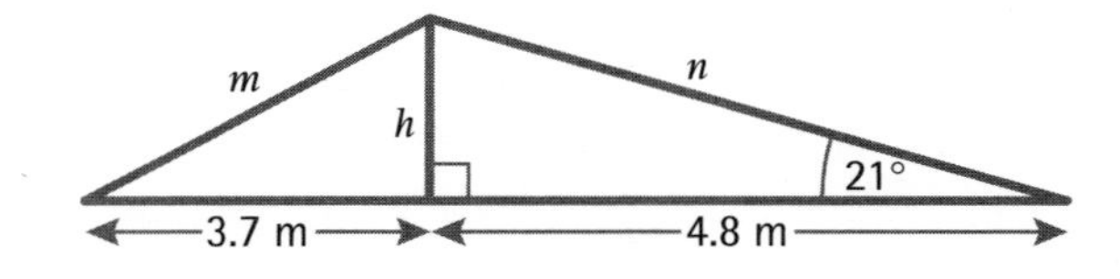

b) Calculate the length of the two slopes of the roof, marked *m* and *n*.

18) Burnbrae Activity Centre plans to build a 'Death Slide'.

The Slide will be fastened to the top of a high tree and to the ground 5.6 metres from the base of the tree. The angle of elevation is 52°.

a) How high is the tree?

b) What length of rope does the Centre need?

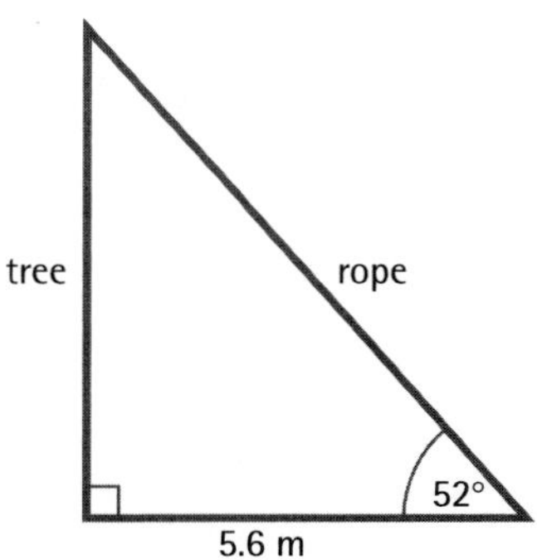

19) (This question ties together the work from this unit, plus some of the work on angles in Unit 2. It is more challenging than the previous questions.)

O is the centre of the circle. AC and BC are tangents to the circle. They meet at the point C.

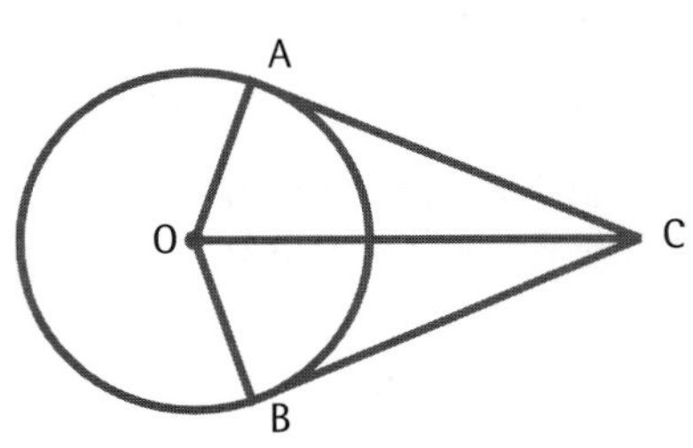

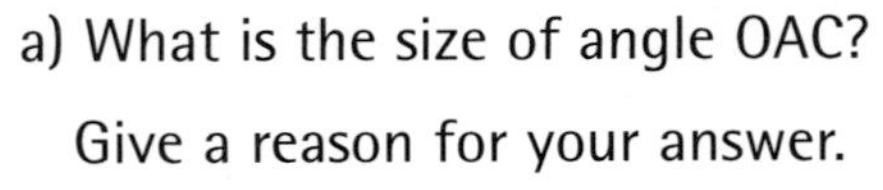

a) What is the size of angle OAC?

Give a reason for your answer.

OC is 14 cm long. Angle ACO is 23°.

b) Calculate the length of the radius OA.

c) Calculate the length of AC.

d) Calculate the size of angle AOB.

Section 5: Algebra

Indices and scientific notation

20) Write the number 86 570 000 000 in scientific notation.

21) The temperature at 3 am was -5°C. By 8 am it had risen by 3°C.

What was the temperature at 8 am?

Patterns

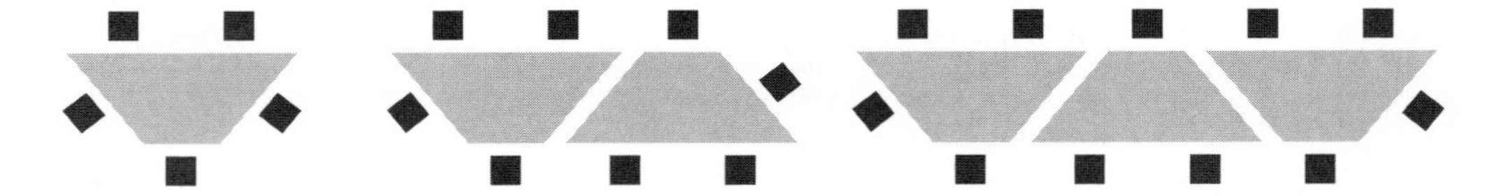

22) Primary classrooms often have trapezium-shaped tables. One table can seat 5 pupils, two tables pushed together seat 8, and so on.

a) The tables are joined together in the same way as in the picture above.

Copy and complete the table below.

Number of tables (T)	1	2	3		8
Number of pupils (P)		8			

b) Write down a formula for the number of pupils, P, who can be seated at a number of linked tables, T.

$P =$

c) A teacher wants to seat 29 pupils, using this layout of tables. How many tables will she need?

Using formulae

23) a) The formula for the surface area of a sphere is $A = 4\pi r^2$ (where A is the surface area and r is the radius).

Find the surface area of a sphere with a radius of 2.5 cm.

b) Find the volume of the same sphere using the formula $V = \frac{4}{3}\pi r^3$ (where V is the volume and r is the radius).

Algebra skills

Exam questions in algebra usually combine the different skills in brackets, simplifying, factorising, equations and inequalities. Questions 24 and 25 are easier examples. Questions 26 to 29 are typical of what you might be asked in the exam.

Remember to check your answers whenever you can.

24) a) Multiply out and simplify $6(v + 3) + 5$

b) Solve algebraically $7p + 3 = 24$

25) a) Solve algebraically $10w - 4 > 66$

b) Factorise fully $8e + 12f$

26) a) Solve algebraically $5p + 2 = 3p + 30$

b) Multiply out and simplify $11(3d - 5) + 15$

27) a) Factorise fully $7rs - 28s$

b) Solve algebraically $8x + 2 < 20 - x$

28) a) Solve algebraically $9y + 2 \leq 2y + 9$

b) Multiply out and simplify $14 + 5(6 - 2y)$

29) a) Factorise fully $25ab + 20b^2$

b) Solve algebraically $8m - 7 = 4m - 6$

Section 5: Using graphs

Gradient

30) Find the gradient of

a) the line a

b) the line b.

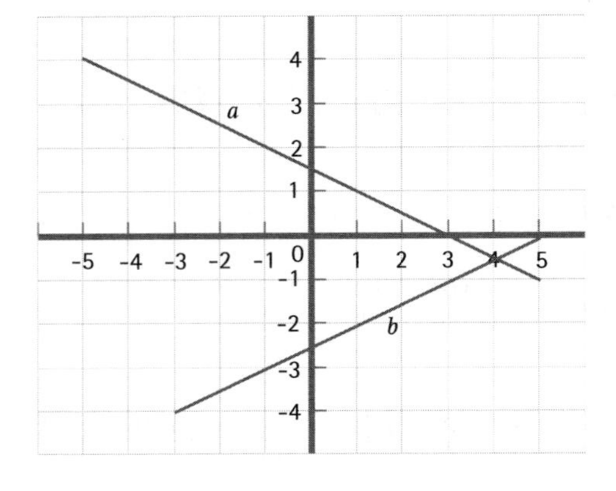

31) Calculate the gradient of each of the ramps.

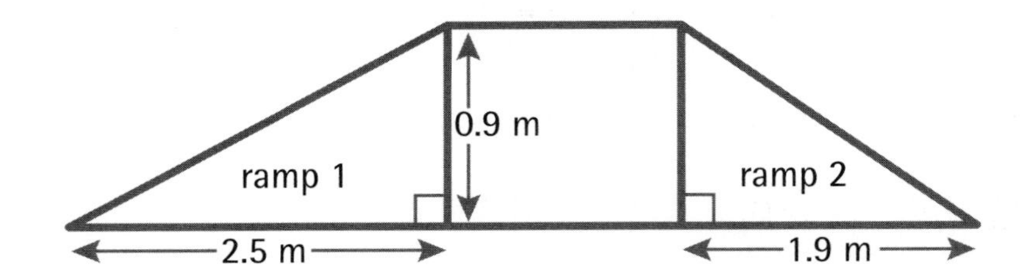

Bar graphs

32) Draw a bar graph (on squared paper) to show the diameter of the following planets. Hint: use a vertical scale in steps of 1000, i.e. 1000, 2000, etc.

Planet	Diameter at the equator in km (to the nearest 100 km)
Mercury	4900
Venus	12 100
Earth	12 800
Mars	6800

Drawing straight line graphs

33) Copy and complete the table below and then draw a graph of $y = 2x - 1$.

x	1	2	3
y			

Answers: exam-style questions

Section 1: Statistics

Probability

1) There are three numbers greater than 5 (6, 7 and 8), so
   P(>5) = $\frac{3}{8}$ = 0.375

Section 2: Using number

Insurance and hire purchase

2) 65 000 ÷ 1000 = 65; 65 x 1.30 = £84.50
3) 2100 ÷ 100 = 21; 21 x 1.65 = £34.65
4) 995 x 20 ÷ 100 = £199 deposit; 995 - 199 = £796; 796 ÷ 10 = £79.60

Ratio and proportion

5) 6 shares of red, so each share weighs 48 ÷ 6 = 8 grams; yellow = 2 shares = 2 x 8 = 16 grams, black = 1 share = 8 grams
6) 30 x 20 = 600; 600 ÷ 15 = 40 minutes
7) 427 ÷ 7 x 4 = 244 pupils

Time

8) 10 (Apr) + 31 (May) + 30 (Jun) + 31 (Jul) + 31 (Aug) + 30 (Sep) = 163 days;
   249 064 ÷ 163 = 1527.9 = 1528 visitors per day

Section 3: Shape and space

Area and perimeter

9) a) angle DEF (angle in a semicircle) = 90°

   b) triangle DEF is isosceles, so angles EDF and EFD are equal

   angle EFD = (180 - 90) ÷ 2 = 45°, so angle EHG = 45° (F angles)

   so angle GHF = 180° - 45° = 135°
10) Area = $\frac{1}{2}$ of (11 x 28) = 154 cm^2
11) Circumference (of two semicircles = 1 whole circle) = $\pi d = \pi \times 38$ = 119.4 m

    Total circumference = 119.4 + 140 + 140 = 399.4 m

Volume

12) Volume = $\pi r^2 h = \pi \times 1.6^2 \times 5.5$ = 44.2 cm^3 = 44.2 ml

Coordinates and symmetry

13) a) and b)

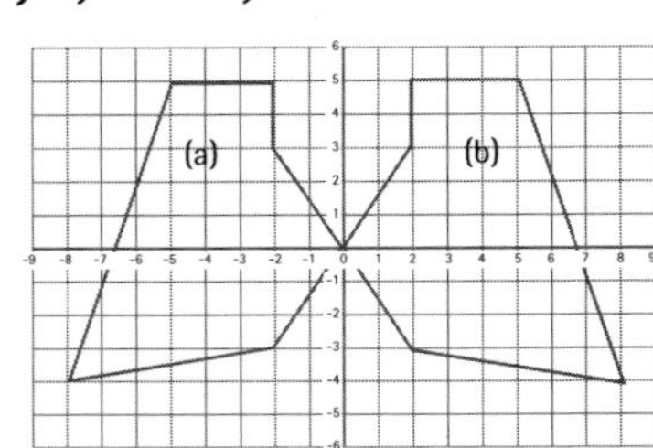

d) (0, 0), (2, 3), (8, 4), (5, -5), (2, -5), (2, -3)

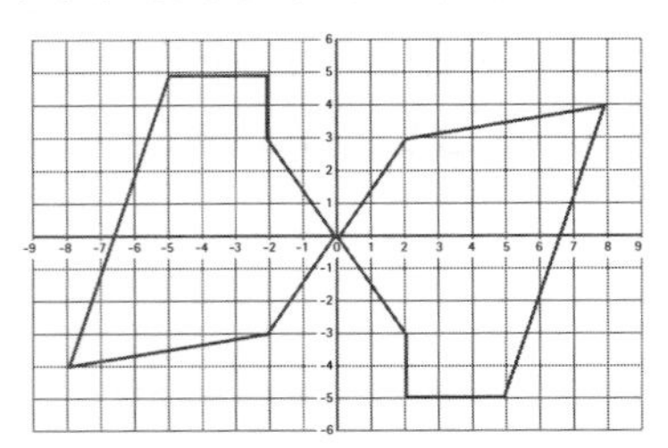

c) (0, 0), (2, 3), (2, 5), (5, 5), (8, -4), (2, -3)

Section 4: Right angled triangles

Pythagoras' theorem

14) Ramp 1: $x^2 = 2.5^2 + 0.9^2 = 6.25 + 0.81 = 7.06$; so $x = \sqrt{7.06} = 2.66$

Ramp 2: $y^2 = 1.9^2 + 0.9^2 = 3.61 + 0.81 = 4.42$; so $x = \sqrt{4.42} = 2.10$

Total length = 2.66 + 2.10 + 1.1 = 5.86 metres

15) $d^2 = 250^2 - 60^2 = 62\,500 - 3600 = 58\,900$; so $d = \sqrt{58\,900} = 242.69$ cm

Trigonometry

16) ✓✓ ✓ ✓ SOH (circled) CAH TOA

$\sin x° = \frac{opp}{hyp} = 60 \div 250 = 0.24$; so $x = 13.9°$

Yes, the ramp is safe because 13.9° is less than 15°.

17) ✓ ✓ ✓✓ SOH CAH TOA (circled)

a) $\tan 21° = \frac{opp}{adj} = h \div 4.8$; $h = 4.8 \times \tan 21°$; so $h = 1.84$ m

b) using Pythagoras' theorem and the answer to a)

$m^2 = 1.84^2 + 3.7^2 = 3.3856 + 13.69 = 17.08$; so $m = \sqrt{17.08} = 4.13$ m

$n^2 = 1.84^2 + 4.8^2 = 3.3856 + 23.04 = 26.43$; so $n = \sqrt{26.43} = 5.14$ m

18) ✓ ✓ ✓✓ SOH CAH TOA (circled)

a) $\tan 52° = \frac{opp}{adj}$ = tree ÷ 5.6; so tree = 5.6 x tan 52° = 7.17 m

b) using Pythagoras' theorem and the answer to a)

length$^2 = 7.17^2 + 5.6^2 = 51.41 + 31.36 = 82.77$; so length = $\sqrt{82.77} = 9.1$ m

19) a) 90° because a radius meets a tangent at a right angle.

b) $\sin 23° = r \div 14$; so $r = 14 \times \sin 23° = 5.47$ cm

c) $AC^2 = 14^2 - 5.47^2 = 196 - 29.92 = 166.08$; $AC = \sqrt{166.08} = 12.89$ cm

d) Angle AOC = (180 - 90 - 23)° = 67°; so angle AOB = 2 x 67° = 134°

Section 5: Algebra

Indices and scientific notation

20) 86 570 000 000 = 8.657×10^{10}

21) $-5 + 3 = -2$°C

Patterns

22) a)

Number of tables (T)	1	2	3		8
Number of pupils (P)	***5***	8	***11***		***26***

b) $P = (3 \times T) + 2$ or $P = 3T + 2$

c) $29 = 3T + 2$; $27 = 3T$; $T = 9$, so the teacher needs 9 tables.

Using formulae

23) a) $A = 4\pi r^2 = 4 \times \pi \times 2.5 \times 2.5 = 78.54$ cm^2

b) $V = \frac{4}{3}\pi r^3 = \frac{4}{3} \times \pi \times 2.5 \times 2.5 \times 2.5 = 65.45$ cm^3

Algebra skills

24) a)

$6(v + 3) + 5$

$= 6v + 18 + 5$

$= 6v + 23$

b)

$7p + 3 = 24$

$7p = 21$

$p = 3$

25) a)

$10w - 4 > 66$

$10w > 70$

$w > 7$

b)

$8e + 12f$

$= 4(2e + 3f)$

26) a)

$5p + 2 = 3p + 30$

$5p = 3p + 28$

$2p = 28$

$p = 14$

b)

$11(3d - 5) + 15$

$= 33d - 55 + 15$

$= 33d - 40$

27) a)

$7rs - 28s$

$= 7s(r - 4)$

b)

$8x + 2 < 20 - x$

$8x < 18 - x$

$9x < 18$

$x < 2$

28) a)

$9y + 2 \leq 2y + 9$

$9y \leq 2y + 7$

$7y \leq 7$

$y \leq 1$

b)

$14 + 5(6 - 2y)$

$= 14 + 30 - 10y$

$= 44 - 10y$

29) a) $25ab + 20b^2$

$= 5b(5a + 4b)$

b) $8m - 7 = 4m - 6$

$8m = 4m + 1$

$4m = 1$

$m = \frac{1}{4}$

Section 6: Using graphs

Gradient

30) a) $-5 \div 10 = -\frac{1}{2}$ or -0.5

b) $4 \div 8 = \frac{1}{2}$ or 0.5

31) Ramp 1: $0.9 \div 2.5 = 0.36$

Ramp 2: $-0.9 \div 1.9 = -0.47$

Bar graphs

32)

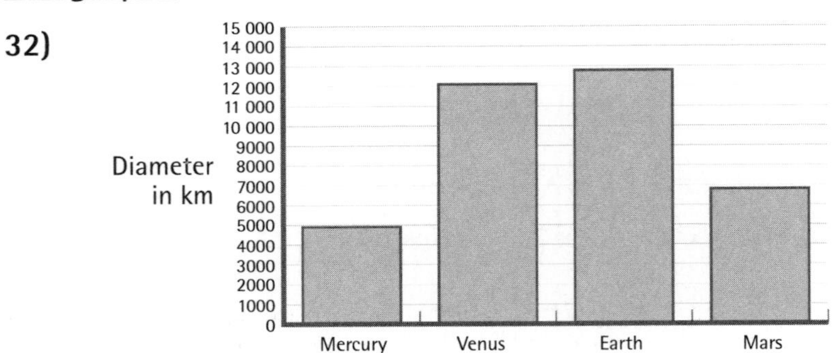

Drawing straight line graphs

33) Table of values:

when $x = 1$, $y = (2 \times 1) - 1 = 2 - 1 = 1$

when $x = 2$, $y = (2 \times 2) - 1 = 4 - 1 = 3$

when $x = 3$, $y = (2 \times 3) - 1 = 6 - 1 = 5$

x	1	2	3
y	1	3	5

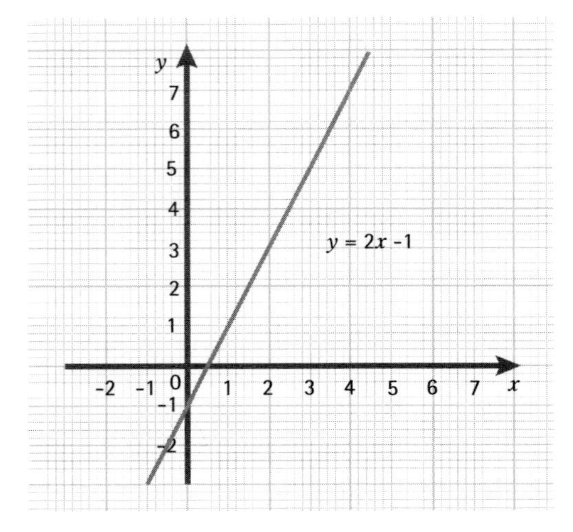